Undeniable

Dialogues on Global Warming

Christopher Keating

Cover design by Don Foote

footeprintz@gmail.com

ISBN: 1500210056
ISBN-13: 978-1500210052

This book is dedicated to all of the climate scientists manning the front lines in the climate war.

Table of Contents

Introduction

I live in rural central Texas and just about anyone in these parts can tell you about how the climate is changing. You can still find a few skeptics, but nearly everyone out here has witnessed it firsthand. Either they make their living off the land or their customers do. Climate change denial is a luxury of the city folks. When you spend years out in the field you don't need someone to tell you things are changing, you have seen it for yourself.

To say climate change makes people emotional would be an understatement. Ask someone for their opinion on the topic and you are likely to get a lively discussion. This is equally true of people that believe in the claims as well as people that don't believe.

The formal name is anthropogenic climate change. Some people call it global warming, others refer to it as climate change. Some people claim those are actually different things, but that is just splitting hairs. The fact is, no matter what you call it, we are all, believers and non-believers, talking about the same thing – human emissions are causing the world to get warmer and this is affecting the climate. This has become the biggest issue of our time for there is no other issue that is discussed by more people worldwide and nothing affects so many people as this one issue.

I am a firm-supporter of the science that shows we have changed the climate. I have been involved, at various levels, in this science for over 30 years and have had the opportunity to review the science, watch as we have expanded our understanding and talked with the leaders in the both the climate science community as well as in the denier community. As a professor of physics and career intelligence officer in the Navy and Navy Reserve I am trained to follow the evidence to where it leads, not where I want it to lead. Make no mistake about it; I want the climate to be normal. I want the pleasant weather of my childhood. Texas in the summer is already hot enough; we don't need to make it hotter. But, that is simply not what the science shows us.

The basis of this book is that the amount of scientific evidence is now so overwhelming that anyone can prove man-made climate change is real and the only way you can believe it isn't is to reject science and logic. I do not say that in an offensive way, I say it as a matter-of-fact. Simply, that is the conclusion the scientific evidence leads us to and to reach any other conclusion would require us to ignore the science.

The evidence, the data and the science to back-up this claim is all available to anyone. The age of the Internet has made more information available to more people than at any time in all of history. Unfortunately, that also means more disinformation is available than at any other time. It becomes the responsibility of the user to realize what is valid and what is snake-oil.

The format of this book follows Galileo's book, *Dialogues Concerning Two New Sciences*. In this book, and others, Galileo had three men discussing the alternate viewpoints of the way the world worked. One man was an advocate of the sciences supported by Galileo. A second man defends the ancient teachings of Aristotle that was the official doctrine of the Catholic Church on the subject. The third man goes back and forth between the two beliefs. Galileo broke new ground with his books by writing them in Italian instead of Latin. That made it possible for the average person on the street to read them and that was something that just wasn't done in those days. Galileo's work got him in trouble with the Church and he had to appear before the Inquisition to explain himself, eventually ending up under house arrest. Proposing a new scientific conclusion back then went against the Church and that was dangerous territory. Today, it is still pretty much the same. Powerful interests want to protect what is theirs.

We have seen this story play out before. In the 1960s, 70s and 80s, an increasing number of scientific studies showed that smoking was hazardous to your health. The tobacco industry funded people to come out and proclaim that was not true and there was no link between smoking and things like lung cancer and emphysema. Of course, eventually, the tobacco companies admitted they knew all along smoking was bad for you and they were just protecting their profit margins. The result was that they had to pay tens of billions of dollars in fines and damages. Today, the very same people tobacco companies hired to deceive the public are now involved in the climate change denier campaign. And, they learned their lessons from the first go round. They are very good at their job and I'm afraid they are winning the battle. Instead of doing something about the problem, we are spending our time debating if it is even real.

If you wish to believe global warming isn't real, isn't caused by manmade emissions, or there is nothing we can do about it, that is your business. You are free to believe in anything you want. I really don't want to get in a debate or argument with you about it. But, if you are interested in learning more about what is going on and want to have an open-minded discussion, then this book is for you. If you are already convinced global warming is real and just want to learn more about it, then I think you will find it helpful. Mostly, this book was written for the people that don't know what to believe. I try to present as claims from both sides of the debate and to show why each is valid, or invalid.

In my book, three friends discuss global warming. In the same manner employed by Galileo, one friend presents the arguments of climate change deniers, one presents the scientific case and the third plays both sides. Their discussion is carried out by email, not only because this provides a bit of updating of the style, but it allowed me to use the subject lines on the emails and allows time for the players to research things before responding.

This also allows me, like Galileo, to present the issue in a debate format. Both sides of the issue are presented and discussed. It is not possible to present all of the claims by the deniers (that would take a very large book) or to present all of the scientific evidence (that would take many volumes of large books), but I tried to present a good cross-section from both.

The three friends - Tom Lawson, Aileen Gulledge and Fred Spencer - do not exist and the exchange of emails did not occur. However, the science is real and many of the discussions in this book come from actual discussions and events that I have experienced over the years. I have included reference notes for anyone that wishes to verify the accuracy of the statements.

I am so convinced of the validity of the science that I have made two cash challenges. The first is the $10,000 Global Warming Skeptic Challenge and the second is the $1000 Scientific Evidence Challenge. The $10,000 Challenge calls for anyone to prove, via the scientific method, that manmade global warming is not real. The $1000 Challenge simply calls for anyone to produce any scientific evidence supporting climate change deniers. Both challenges are open to anyone over 18 and have no entry fee. Entries can be submitted at **DialoguesOnGlobalWarming.blogspot.com**. I am the sole judge (it's my money, after all), but all entries and my comments will be made public. If I fail to judge fairly it will be there for everyone to see. Plus, since there is no entry fee, anyone can submit an entry, wait for my comments, adjust their claim and then submit it again. The deniers claim that it is all cut and dry, so they should not have any difficulty with the challenge. I have presented the scientific argument for anyone to see. Let them put forth a scientific argument to support their claims.

I'm waiting.

Hopefully, my *Dialogues* won't get me before the Inquisition, but I am prepared to stand by the science if it does.

The Earth still moves.

Christopher Keating
Mason, Texas
April 2014

"The great thing about science is that it's true whether you believe it or not"
Astrophysicist Neil deGrasse Tyson

Chapter 1
Why we should care

After a long winter, three friends decide to visit together. Aileen Gulledge, a stay-at-home mom; Fred Spencer, a successful business man; and Tom Lawson, a professor of geophysics, share lunch one spring day.

Fred:
2013 has been a pretty good year, so far. After all, we managed to survive the end of the world in 2012.

Aileen:
It's funny how many of the doomsayers really didn't have anything to say when December 22 came along and the world was still here. The silence was deafening.

Tom:
Sometimes reality wins, at least for a while. But there is always something else waiting in the wings.

Aileen:
Like what, for example?

Tom:

Climate change comes to mind. I constantly meet people who deny that it's real. I had some students asking about it just today. And, of course, there is plenty about it in the media.

Fred:

Don't you get upset having to deal with all of these people and their claims?

Tom:

Sometimes they wear me down, but mostly I get use to it. But, others don't. There is a well organized and financed effort to discredit any research into climate change and to attack researchers involved in it. Some of those guys really get targeted. I remember reading about how Michael Mann said anyone supporting the climate-change debate is immediately attacked by professional climate-change deniers.[1] Mann is the scientist that originated a graph showing temperature levels over the long-term which has become known as the 'hockey stick' because it resembles that shape. A very vocal group of people feel they have to personally attack Mann because they don't like the science.[2] In fact, scientists that publish papers on climate change are subjected to attacks and even have their families threatened. Ben Santer at Lawrence Livermore National Labs had someone leave a dead rat on his doorstep one night. Kenneth Trenberth of the IPCC once said he received over 90 pages of particularly nasty, abusive and hateful emails.[3] I spoke to one climatologist who told me he had to hire a bodyguard when he gives lectures to protect him from physical attacks.[4] It has gotten so bad the board of directors of the American Association for the Advancement of Science issued a statement that they were "deeply concerned" about attacks on climate scientists.[5]

Some scientists take all of this in stride as being part of the business. Unfortunately, these kinds of organized attacks lead others to withdraw from any kind of public debate. This is really bad because it can have a chilling effect on research and the efforts to educate the public about what is being found.[6]

Fred:

This is reason enough for everyone to care about climate change. Clearly, there are groups out there that don't want the public to know what's going on. My first thought is, what are they afraid of? Why do they want to hide this information from the public? I can't believe they are working in my best interest.

Tom:

No, they really are not working in the best interest of the average person. Many of these kinds of attacks are mostly organized by groups funded by industries trying to protect their special interests and are definitely engaged in a disinformation campaign on global warming. They bus people in for organized protests to give the appearance that it is a public reaction.[7] They cannot argue the science, so they are arguing the public relations. They act like lawyers instead of scientists, arguing about every word instead of dealing with the science. Take the 2009 ClimateGate for example, where stolen emails were released. This led to accusations of data tampering and other wrongdoing. Every review has shown there was no misconduct on the part of the scientists that sent those emails.[8] But, the deniers love to point at those emails to support their claims that global warming is a hoax and those emails certainly damaged the public perception of what is going on. Tell me the people responsible for that did not intend that result.

The disinformation campaign is very well funded and very well organized. It is also very effective. Just look at the results of polls to see what I mean. Surveys show 97% of scientists in climate studies believe in manmade climate change. But, only 64% of people in the U.S. think the world is warming and only 47% believe it is due to human activity. Even more telling, only 39% of the public believe most scientists think global warming is occurring and 40% believe there is a lot of disagreement among scientists about it. What is really interesting is that among the most skeptical group of the public only 5% believe that scientists agree. Obviously, this all flies directly in the face of the fact that 97% of climate scientists are convinced.[9]

The news media is playing right into this. They portray the issue as a controversy then give deniers, even the craziest of them, equal air time. This gives the impression that the deniers are somehow credible.

Something like this doesn't happen by chance. There are people working hard to make the public believe this. I am very bothered by these organized campaigns. They are something that we all should care about. It undermines our entire scientific infrastructure.

Aileen:

The old 'attack the messenger' strategy. When you can't prove the research wrong, intimidate the messengers. I think that may be the best evidence that claims about global warming are accurate. If they weren't, there wouldn't be such a strong reaction from the deniers.

Fred:

Yes, too bad it's so effective. People are obviously denying climate change for reasons other than the science.

Tom:

And, we've seen this before. Remember how hard the cigarette companies worked to deny smoking causes cancer? Any guesses about who is behind all of these attacks on climate scientists? It is actually the very same people, and they learned from their mistakes in the tobacco campaign.

The American Petroleum Institute, the API, is behind a lot of it. An organization known as The Heartland Institute is also funding a lot of deniers.[10] They get a lot of donations from major corporations that might be at risk from climate legislation, including one anonymous donor that is responsible for about 20% of their funds and is suspected of being Saudi Arabia. In turn, they fund climate change deniers and work to keep the media from reporting the message of authentic climate scientists. Some of the companies donating to them are Microsoft, Koch Industries, Altria and RJR Tobacco. This campaign they are running is very effective.

Aileen:

It is interesting to see the tobacco industry involved in the denial of science after their experience with shutting down the science on cigarettes and cancer. When are they going to figure out that the truth will eventually come out and it will cost them in the end?

What about fossil fuel companies?

Tom:

Several of them are involved in funding the deniers. I mentioned the API and they are funded by the fossil fuel industry. A group

12

called The Carbon Brief studied over 900 published papers denying climate change and found that 9 out of every 10 climate change deniers had a link to ExxonMobil.[11] This is strange because Exxon has publicly endorsed the idea of manmade climate change and has even committed to investing billions of dollars to develop Arctic oilfields, something that would not be possible without global warming. But, they seem to think they can have more influence on legislation affecting them by funding the deniers.

Fred:

I just have to think that is really silly of them. If it was my company I would come out supporting the science and come off as the white knight. It would be a gigantic public relations coup for some company. And then people will get in their cars tomorrow and fill up the gas tank and life would go on. The difference is that my company would look like a hero instead of a villain.

Aileen:

Too bad they don't think like you do, Fred. It would make things a lot easier.

Fred:

This is another good reason why we should all care about climate change. It is providing another opportunity for the rich and powerful to make decisions for all of us that will affect our lives. And, does anyone really think companies like Exxon or the coal companies or Saudi Arabia care about our well being? I, for one, do not feel better knowing that these companies funding climate change deniers are working with the world governments to influence legislation and policy. It is not to our advantage.

Tom:
They aren't the only ones responsible. There is a pretty strong anti-science crowd out there and they are well organized. There is positively no amount of evidence or argument that you can produce that will succeed in changing their minds. Like you said, Fred, they are rejecting the science for other reasons.

The whole climate change debate has become extremely politicized. People support one side or the other much like they support their favorite sports team. People who object to one party and support another will take the same stand on climate change as their party. Someone doesn't like Al Gore or Barack Obama, two political figures that publicly endorse the idea of man made climate change, so they reject climate change rather than take the chance they are seen agreeing with these people they don't like. Accepting the idea of climate change, to them, is the same as accepting those political figures. Not only is this wrong, but it would be equally wrong to accept the idea simply because figures you like endorse it. Really, people have to look

14

at the science and go where it leads.

Fred:

I am in no way a liberal Democrat, but I believe in global warming. The science is too overwhelming. I've followed it in the news and read articles about it in science magazines. It is a big part of business today. You really have to be able to take it into account when making business plans. There's really no doubt in my mind that we are changing the climate.

Aileen:

I agree. I'm not sure just how bad it is or how much we are responsible for the change, but I'm sure we are responsible for at least most of it, if not all of it.

Tom:

Actually, the science is so overwhelming that anyone can prove global warming is real. You no longer have to be a scientist. In fact, I believe the only way you can be a global warming denier is to reject science. Conversations with global warming deniers can be really strange. I have found that no amount of evidence or logic will ever persuade them that they are wrong.

I'm reminded of a quote I read by E. M. Butler regarding supernatural phenomena. She said that "humanity tends to believe in the teeth of the evidence, or disbelieve in spite of the evidence, but never believes because of the evidence."[12] Supernatural phenomena – global warming denial. There's really not much difference in my mind. Neither has any basis in reality or science.

But, that is the life I have chosen for myself, so I keep trying in the hope that I will make a difference for some.

Fred:
Why do you even try to get them to change their minds? They have also made their choices in life. Why not just let them believe in what they want to believe in and go your merry way?

Tom:
Are you asking me why should I, or anyone else, spend my time arguing with climate change deniers? The short answer is that we shouldn't. They have their right to believe in what they want to. I use to try and change their minds. I even had a reputation about it. Now, I simply ask two questions: "Is there anything I can say or do to change your mind?"; and "Do you think there is anything you can say or do that will make me change my mind?" If the answer to those is 'No', then I see no reason to go any further. I have never had anyone say 'Yes' to either question.

I would like to think that I would say 'Yes' if someone asked me that first question. I like to think that I base my beliefs on the scientific evidence and logical thinking. If someone were to produce evidence that manmade global warming is not real, then I would change my beliefs. That is how strongly I believe in the scientific method. Others, not so much.

Personally, I revel in dealing with the real world, but I've known lots of people that are just overwhelmed by it. I enjoy the obstacles, the challenges, the opportunities, the uncertainties, the twists. I love to wonder about what's around the next bend in the road. I love even more going around the bend and finding out. I can't look at a hill without wanting to go over it and see what's on the other side. Sit me down in a field and I'll be turning stones over to see what's underneath. Isn't life grand?

But, not everyone sees it that way. So, they deny the Nazis liquidated millions because it is too horrible to contemplate. They believe there is a face on Mars because that would mean there is a great civilization out there that put it there and they may provide us with answers. They deny we landed on the Moon because the concept is just so big to them they can't deal with it, so they simply deny it ever happened. These are people that look at the bend in the road as a threat, something to worry about. Hills not only hide what's on the other side from them, but hide them from whatever may lie on the other side. And, these people will never, ever, turn a rock over just for the fun of it. The world is a threat and they are looking for a safe haven.

17

Maybe I'm wrong, but that's the way I see it. There's not much you can do for these people, they have rejected all logic and all science and nothing you can do or say will ever change their minds. I don't mean to say that in a mean way, I am just stating the facts as I see them.

So, why do I bother? Because, maybe I can prevent someone else from denying global warming. Maybe, I can get through to someone that has yet to make up their mind and convince them to check the facts and to think for themselves.

Some people say I'm wasting my time. To quote Thomas Paine "To argue with those who have renounced the use and authority of reason, and whose philosophy consists in holding humanity in contempt, is as futile as to administer medicine to the dead, or endeavoring to convert an atheist with scripture."[13]

Maybe Paine was right, but those that renounce the use and authority of reason today were at one time someone that could have gone down a different path. I may not be able to help the deniers and skeptics of today, but maybe I can help keep someone from becoming one of the deniers or skeptics of tomorrow.

It's all part of the job. I actually enjoy hearing all of these claims, in a perverse sort of way. First, people are going to ask me about some of the claims they make and I want to be prepared. But also, it's like playing a game: How many errors can you find? It helps keep the mind sharp.

Aileen:

Like doing crossword puzzles or Sudoku?

Tom:

Exactly! My favorites are usually the ones that claim to prove Einstein wrong. I recall how one in particular started out saying he had been sitting on a mountainside in northern Thailand when he had this revelation about how Einstein's theory of relativity was wrong.[14] I thought, 'Well, that explains it! All of this time I've been studying and working hard when I should have been sitting on a mountainside in northern Thailand! Boy! I wish I had known that before I started grad school!'

Fred:

Would have saved you a bunch of work, wouldn't it?

Aileen:

Not to mention it sounds like it might be a lot of fun.

Tom:

The thought has occurred to me at times that it might not be too late. Just for the record, Einstein proposed two theories of relativity, special relativity in 1905 and general relativity in

1916. These two theories have been put to extremely rigorous scientific testing and have passed every single test.

Fred:

Don't worry. I don't think we're about to head off to Thailand. Hey! That could be a new catch-phrase for someone making anti-science claims, 'That guy is heading off to Thailand!'

Tom:

Glad to hear you're not heading off to Thailand. Anyway, when I hear someone claiming to prove Einstein wrong my warning senses all start tingling. The game is afoot!

Fred:

Still, there are a lot of people that deny global warming. That is a lot of people to be rejecting science.

Aileen:

People seem more likely to reject science when things are rough. They might be worried that policies will harm the economy or they may be afraid that policies to address the problem will limit individual freedoms or the free market. There may just be a limit to how much people can worry about at one time.

Fred:

According to one study I read, our list of concerns, in order, are the economy, jobs, terrorism, Social Security, education, energy, Medicare, healthcare, deficit reduction, health insurance, helping the poor, crime, moral decline, the military, tax cuts, the environment, immigration lobbyists, trade policy and global warming. It is just not something people are concerned about.[15] One of the reasons for that is because scientists have not shown to people that this is something they need to worry about today. Instead of telling people about the environmental impact global warming will have on our grandchildren, they should be telling people about the economic costs it is having right now. If the economy and jobs are the top two things people are concerned about, show them how climate change affects the economy and jobs. If you show the unemployed the reason they are unemployed is because of the effects of climate change then they will take it much more seriously.

Tom:
Did you really just name all of that off the top of your head?

Fred:
Yes, I did. You do it all the time, so don't be so surprised when I do it.

Aileen:
Fred is right. I think part of the problem is that a lot of the statements scientists make sound like these doomsday

predictions and people have a hard time with that. It's easier to just reject the statements than to believe doomsday is coming. We all already have problems right in our face to worry about. Many people may be asking why we should care about climate change that won't happen for years and years, if at all. They are more concerned with how to pay the bills that are sitting in front of them.

Fred:

I also think the idea of manmade climate change is intimidating to a lot of people. That is the kind of thing that can make people lie awake at night. They deal with it by just denying the idea. Most people avoid extremely unpleasant news until they no longer have a choice.

Aileen:

Maybe another reason people prefer to ignore the idea is that deep down they think that the only solution will require a new way of living and thinking. Change is difficult at best and having to deny ourselves something that makes us comfortable is even harder. Are we going to have to give up our cars and air conditioning? If they just deny the idea of climate change then they don't have to worry about the changes it will bring.

Fred:

What I think is missing from our conversation are real, solid reasons for people to really care about global warming today. We mentioned a few, but those are things people find a way to live with. Powerful people have always been the ones to make the decisions and the average person just has to learn to live with it. There is really nothing new there.

And, it seems like the climate scientists are always talking about what is going to be happening decades or centuries from now. That sounds like it isn't anything we need to worry about today.

Tom, can you give us some real reasons why a typical person should be concerned with climate change today? I mean, things that are happening to them right here and now, not something that may, or may not happen 50 years from now.

Tom:
You both make good points. The scientific community has to get better at communicating with the public. And, we really do need to care about it today. Climate change isn't something that is going to happen next century, it's already happening and it's affecting everyone's lives right now. Let me give an example of something going on right now.

What people really don't get is that climate change comes down to water management. Droughts and floods are going to be worse as the world warms up, as are storms. This is already affecting water distribution all over the globe. In some areas, reservoirs are going dry and crops, livestock and even forests are dying due to the lack of water. Other areas are getting too much water and are experiencing massive floods resulting in billions of dollars in damages. Water management is changing due to climate change and is something that is already affecting nearly everyone on the planet, one way or another.

Aileen:

That's a pretty good reason for people to be concerned about climate change right now. How about another?

Tom:

The expenses from climate change are going to be passed down to the consumer, just as all expenses always are. When food is more expensive, the cost is passed on. When storms result in more insurance claims, the insurance companies will pass it on. Hurricane Sandy caused $75 billion in damages.[16] That comes out to about $250 for every person in the United States. On average, a family of four will be stuck with a bill for $1000 from Hurricane Sandy. I know there is an argument about climate change's role in this case, but the evidence is mounting that Arctic warming was a major factor at work.[17] We might not be completely able to point at a specific storm and say just how much climate change is responsible, but we can look at that total picture and see there is an overall change due to global warming.

So, we know that climate change is, right now, resulting in more severe weather and the cost of that weather is being passed on.

Aileen:

Not to mention the extra money I have to spend to pay the air conditioning bill every summer. It seems like it keeps going up every year. Hitting the pocket book is something we all are concerned about. If climate change is costing us money then I would think people would be interested.

Tom:

Something else hitting close to home is food. One of the real hazards of climate change is food production. We have seen that climate change in the past has disrupted food supplies resulting in mass famines that have killed millions and even brought down cultures. Remember that we saw drought was one of the factors for the fall of the Mayan civilization. Drought is also attributed as a factor causing the fall of the Ming Dynasty in China. But, drought isn't the only concern. Incredible floods in Europe in the 14th century resulted in massive crop losses and the collapse of the economy in many nations. What we are experiencing today is greater than anything any of those previous cultures ever experienced.[18]

Aileen:

That sounds like doomsday scenarios, Tom.

Tom:

Let me give you something on topic but less dire. There are some studies showing increased food production is possible in the short term, but we really aren't seeing that as much as expected. While the warmer average temperature will help with increased production, other factors work against it. Increased droughts, heat waves, hail storms and other forms of severe weather all destroy crops. So, even if climate change results in new areas for growing crops it works at the same time to destroy them.[19]

Food production between 1980 and 2008 has not increased as much as it could have without rising temperatures. Corn and wheat crops have been getting larger, but not as much as they could have. Because of rising temperatures, corn production is down 3.8 percent and wheat production is down 5.5. It is estimated that this has caused the price of corn to go up 6.4 percent and wheat to go up 18.9 percent.[20]

Fred:

That isn't the end of the world, but it takes money right out of the average person's pocket.

Tom:

One of the things we are seeing is increased droughts, not the end of the world type, but still big ones. For instance, the 2011 drought in Kenya left 2 million people and 20 million heads of livestock in emergency need of food and water. Drought isn't new to that region, but they had already experienced a severe drought just two years earlier.[21] As much as that may be a terrible tragedy, it isn't the end of the world.

Another example comes from Australia, which went through a massive, 10-year long drought that resulted in widespread wild fires and dust storms. The size of sheep herds dropped by 50 percent and rice and cotton production collapsed in some years. The drought affected tens of thousands of farm families across Australia before it ended in 2010. And, as is often the case, it ended with torrential rains and flooding. The evidence is piling up that climate change may have been responsible for that drought. At the very least, it severely aggravated it. The American southwest may be in a similar situation with long-term rainfall decreasing across the area.[22] Again, not doomsday, but still a terrible tragedy. And, as I said before, the costs from these events are being passed on to consumers.

Fred:
That is a good example of things that we need to be concerned about right now, but aren't the end of the world. If we take steps we can get through that, but we need to get busy.

Tom:

Adapting is the big key and is why we need to be concerned today. Africa, in particular, can expect to see more frequent and more severe droughts in years to come. They need to start adapting now.[23]

Weather in general is getting more extreme and we can expect more violent and extreme weather events.[24] According to data on weather events, floods, droughts, wildfires, windstorms, blizzards and tornadoes are all becoming more frequent and more severe.[25] According to NOAA, the U.S. had at least 12 weather disasters in 2011, each of which caused more than $1 billion.[26] That comes out to at least $40 for every person in the U.S., or about $120 for a family of four. This is happening today, not some distant point in the future.

Aileen:

These are still big, scary changes. Can you list some changes that aren't so scary, but still going on right now?

Tom:

Allergies are getting worse. Plants are producing pollen earlier and making more of it so people are having worse allergies.[27]

Fred:

I know some people would consider that to be a doomsday scenario. And, they would probably confirm that their allergies have been getting worse.

Tom:

We can expect to see more and more changes to the ecosystems of the world and we live in those ecosystems which mean there are already lots of changes that are affecting us right now. This is already occurring, but it will happen more quickly in the future. Almost half of the world's land area can expect to see its ecosystem change. Tundra, forests, grasslands will all turn into something different. Again, this isn't the end of the world, but it will have a big impact and we will need to adapt.[28]

Fred:

It's pretty obvious that there will be lots of changes occurring and some of it is already happening. We really do need to make plans to adapt. There will be winners and losers. The winners will be the ones that make plans and the losers will be the ones that don't. Kind of like the Eastman Kodak company. They were at the top of their industry but didn't adapt when digital cameras came along. They are even the ones that invented the digital camera.

Aileen:

The problem with your statements, Tom, is that they don't prove global warming is real. These are all good reasons for every one of us to be concerned with global warming, but if it isn't real, then these problems won't occur. I think most people would be concerned about these things you listed, but if they don't believe the climate is changing, then they don't believe these things will ever be a problem.

Fred:

She makes a good point, Tom. We need to prove global warming is real. And, I don't mean prove it to the scientists, but prove it to the person in the street.

Tom:

It sounds like we have a job on our hands. Where do you want to start?

Of all the animals, man is the only one that lies.

Mark Twain

Chapter 2
Skeptics and Deniers

Aileen:

I think if we are going to get into a discussion on the issues we should start with the people involved. Why don't you tell us about them?

Tom:

I actually break the people who don't believe in global warming into two groups, the skeptics and the deniers. Deniers are people that are actively working to fight against global warming and trying to convince people it isn't valid. Many of them are getting paid, directly or indirectly, by the fossil fuel industry. Skeptics are the average people that don't believe in global warming, but aren't out campaigning, either. Most skeptics are simply sucked in the by arguments of the deniers.

I don't want to sound like I'm taking a stand here, even though I am, but the arguments the deniers make are devoid of science.

For instance, I remember a TV interview with a guy named Joe Bastardi, who claimed global warming is impossible because it would violate the first law of thermodynamics.[29] The first law says you can't create or destroy energy, therefore, according to him; putting CO_2 in the air can't create energy and can't warm the planet. It was so bad that any undergraduate physics major could punch it full of holes, yet there he was on TV and they were taking him so seriously. Then there was this article[30] I saw about this same guy leaving Accuweather. Officially, he resigned, but I believe he was forced out because he kept making so many anti-scientific claims.

Fred:
Even I can see the problem in his claim. Global warming theories don't claim CO_2 is adding energy to the atmosphere, they say CO_2 traps energy that would otherwise be going off into space. That's pretty lame of that guy. And, I might add that it is pretty lame of the TV network to give him air time.

Tom:
The principle aim of the deniers is to make everyone believe we don't need to do anything about global warming. They are in the business of protecting special interests that stand to lose money if global warming is real and they have an arsenal of tools at their disposal. They use false arguments, outright lies, cherry-picked data, half-truths and carefully selected truths, frequently taken out of context. As long as they can keep the controversy alive little action will be taken to address the problem and we

will not be doing those things necessary to adapt to the new situation.

Adapting is the key. Climate change is here to stay. If we completely stopped producing greenhouse gases today there is already enough in the atmosphere that it would take centuries to get back to pre-industrial levels. It is typically thought it will take 100-200 years for the environment to remove all of the gases that are currently in the atmosphere but some models have shown it may take well over 3000 years.[31] The amount of CO_2 in the atmosphere of today is greater than at any time in at least the last 800,000 years.[32] In fact, the amount of carbon dioxide in the atmosphere has increased by 36% in just the last 150 years and half of that was added in the last 30 years. But, adapting means someone loses out. People in the business of selling stuff that produces greenhouse gases will lose money.

Fred:

But, the production of greenhouse gases is part of life and industry. Just by living we will produce greenhouse gases. As the world population increases and becomes more industrialized we will continue to produce those gases. We need to learn how to control our emissions and how to adapt to the changes we are making to our planet. But, we are not stopping the production of greenhouse gases. Not today, probably not ever. Realistically, the best we can hope for is to reduce them and keep them under some acceptable limit.

Aileen:

And yet, there are those that, for what ever reason, refuse to believe. We have seen several reasons why we should all care, but so many people just ignore the whole situation.

Tom:

It's pretty mysterious. People need to understand that taking no action is actually making a decision, the decision to do nothing and let events play out as they will. And, that's a bad decision because things will not play out to our advantage if we leave them the way they are. The scientific evidence is so overwhelming that anyone with an open mind can do just a little bit of homework and see that the theory is valid. I would bet that any person of average intelligence could prove the validity of manmade global climate change with the information that is available.

Fred:

That sounds like the wine talking, Tom.

Tom:

No, not at all. I really do believe there is so much information available now that anyone can prove global warming is real, if they are willing to do their homework. You don't need to be a scientist, you just need an open mind.

Fred:

What do you mean by 'open mind?'

Tom:

Someone that is willing to apply the scientific method and live by the results. Universally, every global warming denier I have ever met or read about has rejected the scientific method on this topic. And, I mean without exception. They will look at the scientific data and just reject it because it doesn't fit their preconceived conclusions. We have been able to show without question that the climate is warming and human activities are the main cause.[33] The only way that you can deny that is by rejecting science.

Aileen:

But, aren't there some scientists out there that are skeptical? Are you really saying they have rejected the scientific method?

Tom:

Sure, why not?

Fred:

Why would a scientist, or anyone else for that matter, reject science?

Tom:

I think most of the time when some average person rejects global warming it's because they have been convinced to do so by the deniers. I consider these people to be victims. It isn't their fault. Very professional salesmen have sold them a bad bill of goods.

There was this time I made a presentation about the Arctic sea ice and showed how the extent of sea ice has dramatically decreased over the last 30 years. After the talk a guy in the audience came to talk to me and ask some questions. His claim was that the radiation from the Fukushima nuclear reactor in Japan was responsible for the melting of the Arctic sea ice and he wanted to know what I thought of that.[34]

First, the Fukushima disaster occurred after the 2011 earthquake and tsunami. The sea ice extent has been declining since at least 1980, long before the earthquake ever occurred. But, here's the bigger issue, the amount of radiation released from Fukushima is trivial on the global scale. Outside of the Fukishima exclusion zone the level of radiation falls way below background radiation levels.

He asked me if I would feel comfortable swimming off the coast of Washington state and I told him I would have no fear at all, at least not due to any radiation. The plume of radiation that is working its way across the Pacific Ocean is harmless and may even be so low that it is undetectable by the time it gets to Washington. No monitors have even detected it yet.

I asked the guy if he knew anyone that smokes. He admitted that he is a smoker himself. I pointed out to him that the amount of radiation he gets every year from smoking a pack a day is more than four times the average total dose as a result of the Three Mile Island accident. A pack a day of cigarettes provides about as much radiation as a mammogram. Yet, he wasn't concerned about this radiation. He wasn't concerned that the radiation from his cigarettes is greater than the amount of radiation he was blaming for the melting ice caps.

This is an example of the problem we have educating the public. This guy made two assumptions, neither of which has any credibility: the radiation from Fukushima was enormously larger than the reality; and this radiation could melt the polar ice cap. He then used these two bad assumptions to come up with an invalid explanation for an observed event in an attempt to find a conclusion to support his beliefs.

Ultimately, the problem was he did not want to believe in global warming. When presented with incontrovertible evidence of what is going on, he wanted to deny global warming so much he came up with a completely unrealistic alternative explanation.

The easy, and obvious, explanation for the observed melting of the Arctic Ocean is global warming. The evidence is overwhelming. And yet, when confronted with something that challenged his beliefs, he rejected all of that evidence and created this fantasy that allowed him to maintain the beliefs he wanted to keep.

And, that is the moral of the story. So many people have reached the conclusion they want to believe in and no matter what the evidence says, they will continue to believe what they want to believe. And, if necessary, they will invent illogical and unrealistic scenarios to allow them to continue the way they are.

It's an uphill fight.

The big deniers are a different matter, though. Some of the deniers are being funded to reject the science and push the denier agenda. For instance, two of the most visible deniers are Craig Idso and Fred Singer, both of whom are funded by The Heartland Institute. They give Idso $11,600 per month and Singer $5000 per month plus expenses.[35]

Other people simply have a hard time accepting the consequences. Scientists are subject to all of the weaknesses other humans are. We like to think we are above emotional decisions, but the record clearly shows we are not. Like you guys pointed out, skeptics are almost always skeptical for other, non-science reasons. Maybe they are opposed to proposals for the government to regulate or control certain things. Maybe they are just scared of the idea that things are spinning out of control. Motivations like these affect scientists just like they affect anyone else. And, don't be fooled. Over 97% of scientists that are active in the field agree that climate change is occurring and it is primarily caused by human activity.[36]

Aileen:
Wow! You almost make scientists sound like they're human!

Fred:
Let's not get carried away, Aileen.

Tom:
Some of them, at least.

There are extremely few scientists in the climate-related sciences that have rejected global warming theories. Nearly every scientist that is on record as being a denier is in some other discipline. Also, many of the 'scientists' you speak about are not formally educated in the sciences or employed as a scientist. You don't need a license to be a 'scientist,' there is no bar association or AMA for scientists. You can just hang your shingle out front and start making claims. In fact, that is exactly what some of these people have done. I get emails from high school students calling themselves 'a scientist.' If some guy retires after spending his life reading science magazines and decides he wants to be a scientist, he can make that claim. Technically, they are, but not in the traditional sense.

Fred:

What about that Nobel Prize winner a couple years ago? He is certainly a scientist in the traditional sense and he quit the professional union he was in because he disputed global warming claims.[37]

Tom:

Yes, Ivar Giaever resigned from the APS, the American Physical Society. I remember him. Part of his complaint was about global warming. But, he also had a complaint about the wording in the society's statement on global warming. The statement stated that 'the evidence is incontrovertible,' and he felt that this precluded any discussion on the issue. Some people have pointed to errors in the IPCC report to support that claim. The fact is, the report is over 3000 pages long and there will always be some errors in a document that large and complicated. But, none of the errors affect the substance and conclusion of the IPCC reports and the reports remain unchallenged in the research community.[38]

I agree with him that there should always be debate and

discussion because that is part of the scientific method, but I believe the wording is correct. Any debate has to involve scientific evidence and be the result of rigorous scientific research. Like I said, I have never seen anything produced by the deniers that approaches that standard.

Really, there can be no rationale scientific conclusion other than manmade emissions are causing global warming. Any debate at this point should be about the details and how we can move forward. Just because we can prove global warming really exists doesn't mean we understand everything about it. In fact, it's just the opposite, there is much we don't know or understand. That part is open to debate but we shouldn't be spending our time arguing about the role of manmade emissions in global warming any more than we should be debating if the Earth orbits the Sun.

Aileen:
But, Tom, the guy is a Nobel-prize winner. Are you still saying this guy has rejected the scientific method? Isn't that harsh?

Tom:
I don't think I'm being harsh. I'm just showing my faith in science and the scientific method. He won the Nobel Prize for his work in quantum solid state physics, how matter works at the sub-atomic level. I would not want to debate him about tunneling in semiconductors. But, he is not a climatologist so I have no

41

problem saying he is wrong on this issue. And, if he applied the scientific method to subjects that he is not expert in I contend he would reach a different conclusion. I believe he is reaching the conclusion he wants, not the conclusion that is supported by the scientific evidence.

At the same time I would point to the skeptic that did a big study to prove there was no global warming and came away believing there was. His name is Richard Muller. He and his group, Berkley Earth, spent two years examining data before reaching the conclusion the climatologists were right.[39] That is the way the scientific method is suppose to work.

The fact is the Earth is absorbing more energy from the Sun than it is sending back out into space.[40] Given that scenario, it is inevitable that the Earth has to be warming. We know that planetary heat is trapped by greenhouse gases. We can see that the amount of greenhouse gases is going up and that this is due to manmade activities. These are scientific facts. The rest is details.

Debating whether or not there is manmade climate change makes no sense. I don't need to get into a debate about whether there is a tooth fairy. It's about the same. And, saying that climate change is coming is like standing in a foot of flood waters and saying a flood is coming because the three-foot flood crest hasn't arrived yet. We are in climate change right now.

Aileen:

But, don't you think it is healthy to question scientists? Just like we should be questioning any voice of authority? There needs to be trust, but shouldn't there also be some skepticism?[41] I think the skepticism and the questioning actually leads to better trust. I think we, the non-scientist public, need to be able to question scientists about topics and receive satisfactory answers to our questions. We shouldn't just blindly accept what we are being told.

Tom:

I agree. And, keep in mind that policy also extends to scientists. I may be expert in my area of study, but I can't be expert in everything. At some point, I have to trust fellow scientists and accept what they are saying. But, as you said, we have to be able to question them and receive good answers. This is what I'm talking about with that guy that resigned from the APS. He is expert in semiconductors, but he is not expert in climatology. He has a right to question what is being said. In fact, as a scientist, I believe he has a duty to question what is being said. But, when he receives answers he doesn't like he can't just ignore them. I believe the evidence is overwhelming and he would have to reject the evidence to reach the conclusion that there is no global warming. Questioning is healthy, but it must be done in a

43

responsible manner. Questioning does not mean you question that the Sun will rise in the east. At some point, you have to accept the science.

Aileen:

I guess it could it go too far. What about these deniers you were talking about? It doesn't sound to me that they are really questioning, responsibly or otherwise, they are outright rejecting what scientists are saying.

Fred:

Yes, that would be a difference between the deniers and a reasonable, questioning individual. I think we need to accept that scientists are experts, but expect them to justify their statements to us. But, when you question someone you have to be prepared to accept the answers. The deniers seem to have decided that they are smarter than the scientists and deny everything that doesn't agree with their preconceived beliefs. The typical climate change denier isn't trying to get an answer. They are trying to corner the scientist and score points.

Tom:

Well said. The problem is that so many people fall victim to the deniers. There are people that hear what the deniers have to say and just accept it without doing any real homework. They just read blogs by other deniers and take what they say as truthful.

It's rather ironic. They accept anything the deniers have to say, but they won't accept anything the scientists have to say.

Much of the debate in the public forum is not healthy or responsible.

Aileen:
I understand. I might not be able to completely understand the science and methods involved, but I can recognize false arguments and valid reasoning by using critical thinking skills.

Tom:
Right! And, of course, don't forget the scientific method. Anyone can check to see if the scientific method is being used. We discussed that a lot last year when we were talking about 2012. You can also question assumptions that are made and how the data was collected. Again, there is enough data and literature out there that anyone can do this. Unfortunately, there are those that ignore the science and go for the fringe claims that have no supporting evidence, or to claims that just don't matter.

Aileen:
Let me ask you something, Tom. You said nothing can convince the deniers they are wrong and that global warming is real. But, what if they are right? What would it take to convince you to change your mind?

Tom:

That's a fair question and I can answer that one. All they have to do is provide the proof via the scientific method that the claims about climate change are not real. I will believe in the science, no matter which way it goes. But, to date, all of the science shows climate change is real and I am going to support the science.

Fred:

I'm intrigued by this claim of yours that anyone can prove global warming. It seems to be counterproductive. You are saying the scientific data is out there for anyone to demonstrate that theories about manmade global climate change are valid. Doesn't it stand to reason that there is also enough data to prove they are false? Aren't you opening the door to everyone in the world to try and do this?

Tom:

You are completely correct. If the theories are wrong, the data is out there to prove it. But, if that is the case, why hasn't anyone done that? And, I don't mean one of the 'violates the first law of thermodynamics' proofs. I mean a valid, rigorous proof that follows the scientific method and is reproducible by other scientists. And, yes, this does open the flood gates for every denier out there. But, the game has to be played that way. Scientists don't have magical powers. Anyone can do science with proper training. Therefore, you reach the conclusion that the whole populace is entitled to try to prove their theories. I am on record as being a firm believer in manmade global climate change. Technically, the term is anthropogenic global climate change. But, if someone could prove, via the scientific method, that the theories are not valid, then I would change my position. You have to go with the science. The problem is the deniers are rejecting science, not using it.

Fred:

You say the data is out there. Where, for instance?

Tom:

A quick Internet search will turn up lots of scientific sites devoted to the topic. Just off the top of my head, I know of the Pew Center on Global Climate Change,[42] the NASA Goddard Institute for Space Studies,[43] NOAA Climate Services,[44] NASA Climate Change[45] and the Intergovernmental Panel on Climate Change, known as the IPCC.[46] RealClimate.org is a good website for information from climate scientists. ThisisClimateChange.org is another. There are lots more, but those come to mind quickly.

Fred:

I'm getting an idea. Why don't we do another dialogues, like we did on 2012? Why not do one on manmade global climate change? The one on 2012 was a lot of fun and I really learned a lot from it. What do you guys think? We still have the list serve and we could use it to keep our dialogues emails separate from our regular every day emails like we did before.

Aileen:

I like it. I agree the 2012 dialogues was a lot of fun. We can use the format of Galileo's *Dialogues Concerning Two New Sciences*[47] again. One of us can be Salviati, arguing for the theory, one can be Simplicio arguing against it, and one of us can be neutral Sagredo.

Fred:

Sounds great. But, Tom, you can't be Salviati again.

Tom:

Why not?

Fred:

You said the information was out there for anyone to prove the theories valid. Let's put it to the test. Aileen or I will be Salviati this time, proving the science, and you be Simplicio, the naysayer. You come up with reasons why global warming isn't happening or it isn't manmade and we have to prove that it is. This way we can show if a non-scientist can prove or disprove the theories.

Tom:

I like it. Although I have to warn you I have lots of stuff to use. I keep all of the emails I get from deniers, and I have a lot of them. Also, I get questions from my skeptical students all the time.

Fred:

That's OK. That is the stuff that the average person on the street hears, so that is the stuff we need to be able to deal with, just like anyone else.

Aileen:

How do you want to go about this?

Fred:

Let's start with what we were talking about, namely how you would go about proving manmade global climate change. I think you have to start with the basics. Is the Earth warming? I know a lot of people still don't believe there is any warming going on. Clearly, if you can't prove the Earth is warming then all the rest is irrelevant.

Aileen:

I think you're right; there are still lots of people that are skeptical global warming is even occurring. There are also people that

49

accept the Earth is warming, but think it is all due to natural climate change. I think the next step should be to prove that, if the planet really is warming, it isn't due to natural causes. If the planet is warming, but not because of the greenhouse effect, then manmade emissions don't matter.

Fred:

Good idea. But then, we are at the real crux of the question. If the Earth is warming and at least some of it is due to the greenhouse effect, are manmade emissions contributing to that effect? There could be an increase in greenhouse gases purely due to natural causes and manmade activities are not doing anything to contribute to the problem. Or, manmade greenhouse gases are contributing, but are insignificant compared to natural causes.

Aileen:

Great! If we can apply the scientific method to each of these questions and the answer is 'Yes' to all three, then we can safely conclude that manmade activities are, at the very least, contributing to global climate change, or global warming. A 'No' answer to any one of these questions would indicate that manmade activities are not causing global climate change.

Tom:

This all sounds like a good start, but then we would need to address the issues the deniers bring up as 'evidence' that there is no climate change. Such as claims that the theories and the data are all wrong. If you have a bad theory and bad data the result you reach is going to be incorrect. We need to examine their claims and show if they have any validity or not.

Fred:
Like the guy that claims it violates the first law of thermodynamics?

Tom:
Yes, but there are other objections. For instance, we know there are natural variations in the climate.[48] How much of any observed climate change is due to natural changes that would occur even if we were not here?

Aileen:
I think we need to talk about what could be done about it, also. There are people that claim global warming is a good thing and we don't have anything to worry about. And, those that say we need to take all sorts of draconian measures. I've heard people saying we need to basically go back to horse and buggy days. I would like to think there are things we can do that aren't quite that extreme.
Fred:
Just for the record, let's review exactly what we mean when we

say we're following the scientific method.

Aileen:

As I recall, the scientific method consists of four steps that repeat over and over: 1. Observation of some phenomena; 2. Forming a hypothesis or theory; 3. Using our hypothesis to make a prediction; 4. Testing the prediction. Then, the process begins again by observing the results of the test and adjusting your hypothesis and making new tests. I also remember you said the definition of a science[49] is anything that follows the scientific method. If it doesn't follow the scientific method, it's not a science. Cut and dried.

Tom:

Right! And, as we saw before, one problem scientists have talking about science with the general public is that scientists use different definitions for some words, like 'hypothesis' and 'theory.' The common definition of 'theory' is a best guess. But, in science a best guess is called a hypothesis. A theory is something that has been tested and passes all tests. A hypothesis can only be elevated to being a theory if it reaches the point of never failing the tests. There are just two grades in the scientific method, 100% and failure. So, in day-to-day language, deniers will say, 'that's just a theory!', but in science, to say something is a theory means that it has been rigorously tested and it has passed all tests.

Aileen:

So, we need to use these steps to prove global warming?

Tom:

Yes, but we don't have to actually do the experiments or collect the data ourselves. We can use data that was obtained and tested under the scientific method that is already available. That is why I say anyone can do this. Scientifically valid data is easily available to anyone that wants to examine the issue themselves. Plus, there are lots of scientific papers and news articles about these results. There are also those deniers you mentioned out there calling the data into question. We can examine those claims to see if there is any validity to them under the scientific method. If the scientists have to conform to it, so do the deniers.

Fred:

Alright, it sounds like we have a plan. We may need to add something along the way, but that won't be a problem.

Tom:

Just one question, are you going to put all of this together into a book again, like you did for our 2012 discussions? I would love to have one on global warming.

Aileen:

Oh, yes, Fred! That was such a nice book you put together. I use it all the time.

Fred:

But, I'll have to type up our conversation tonight from memory. And, since I'm doing the work, I get to be the hero.

Tom:

Of course. We wouldn't want it any other way.

To mistrust science and deny the validity of the scientific method is to resign your job as a human. You'd better go look for work as a plant or wild animal.

P. J. O'Rourke

Chapter 3
Is the planet warming?

From: Tom.Lawson@university.edu
To: Dialogue@email.com
Subj: Kick off
Date: Tuesday, May 7, 2013 8:27:52 AM

Have you two decided who will be Salviati, arguing for the theory, and which one will be neutral Sagredo?

From: Aileen.Gulledge@email.com
To: Dialogue@email.com
Subj: Re: Kick off
Date: Tuesday, May 7, 2013 12:58:01 PM

We talked it over. I'm going to be Salviati this time. I really like to idea of using the emails. Unlike you, Tom, I don't have this information in my head and I'll need time to research the issues and get back to you guys. Fred will be Sagredo and play both sides.

From: Fred.Spencer@email.com
To: Dialogue@email.com
Subj: Scientific method

Date: Tuesday, May 7,2013 3:14:40 PM

I'll start doing that right away by pointing out that the scientific method cannot 'prove' a theory, only disprove it. Therefore, it is not possible to conclusively prove that anthropogenic emissions are resulting in global warming. Instead, what we can do is show this theory passes all tests that the scientific method puts it to. Is that right, Tom?

From: Tom.Lawson@university.edu
To: Dialogue@email.com
Subj: Hypothesis
Date: Tuesday, May 7, 2013 8:54:01 PM

Yes. We can put a hypothesis to the test and it will either pass the test or fail. A failure shows there are flaws in the hypothesis and more work needs to be done. A pass only means it passed this particular test but doesn't rule out that it might fail some future test. Eventually, we can test a hypothesis enough that we can accept it as theory, but even then we continue to test it.

Of course, deniers like to use this fact as evidence that climate change science is invalid. They will say that the science may change at some point in the future and that is a convenient half-truth. There will always be advances in science, but any advance must include what we have already learned and discovered. A hypothesis failing a test does not necessarily mean it is invalid, it just means more work is needed.

From: Aileen.Gulledge@email.com
To: Dialogue@email.com
Subj: Step 1
Date: Wednesday, May 8, 2013 2:14:12 PM

So, we need to show that this theory about manmade emissions causing global warming can pass all current tests. I've been doing some homework and, being the one arguing for the theory, I'm willing to say I think we can do this.

We agreed the first step in our plan of action is to see if the Earth is even warming. If the evidence does not show the Earth is warming then we can conclude there is no evidence that manmade emissions are causing global climate change and we don't need to go any further. But, if we find that the Earth really is warming it does not mean manmade emissions are responsible. That will require more work. Are we all okay with that statement?

From: Fred.Spencer@email.com
To: Dialogue@email.com
Subj: Re: Step 1
Date: Wednesday, May 8, 2013 4:07:55 PM

I'm good.

From: Tom.Lawson@university.edu
To: Dialogue@email.com
Subj: Re: Step 1
Date: Wednesday, May 8, 2013 6:38:16 PM

I'm okay with it.

From: Aileen.Gulledge@email.com
To: Dialogue@email.com
Subj: Indicators
Date: Friday, May 10, 2013 4:07:55 PM

There are a number of indicators we can check to see if the Earth is warming. The data on each of these is available on the Internet for anyone to check. Before we even get into the scientific data we can see lots of evidence of global warming. For instance, let's look at winters. It may sound strange, but winter storms are actually proof of global warming.

From: Fred.Spencer@email.com
To: Dialogue@email.com
Subj: Re: Indicators
Date: Friday, May 10, 2013 8:12:08 PM
How's that?

From: Aileen.Gulledge@email.com
To: Dialogue@email.com
Subj: Winter weather
Date: Saturday, May 11, 2013 1:16:40 PM

Winter is due to the Earth's tilt relative to the Sun which causes the poles to spend six months in darkness. Try spending six months without sunlight and see how cold it gets. Winters are getting shorter and milder, but they won't go away completely.

The thing to remember is that Earth is still tilted on its axis and the four seasons will keep happening. We will still have equinoxes in the spring and fall and solstices in the winter and summer. What this means is that the Polar Regions will have long periods of darkness or low light. The exact amount depends on how far you go away from the pole. This means the air over the Arctic Ocean will sit in darkness for a long period of time and it will get very cold. Global warming is not going to stop that.

But, what global warming does is to provide the energy needed to move this gigantic mass of cold air. Left to itself, the cold air will just sit over the Arctic until spring comes along to warm it up. But, currents in the atmosphere will not leave this cold air mass to itself and never has. The atmospheric currents work to move masses of air around, including the cold air mass over the Polar Regions. This, of course, has been the way things work for millennia, at least. What has changed, though, is how much energy there is to move this cold air mass. So, if there is more energy to move the air mass out of the Arctic, we would expect to see some increase in winter storms, even with a winter that is milder over all.

To move trillions of tons of air from the polar regions to lower latitudes takes an awesome amount of energy. The atmosphere is a big heat engine and needs energy to move air masses around.

The more energy there is in the atmosphere, the stronger the air currents will be, which means more air will be moved around to greater extent and at higher speeds. As the global temperature rises, the amount of energy available for storms will also rise. Global warming should result in shorter winters that begin later and end earlier, with more moderate average temperatures, but with more severe storms.[50]

And, this is precisely what is being observed. The record shows that storms are getting more energetic and more frequent while the average temperature is going up at the same time. That means the atmosphere actually has more energy in it today than it did in the past, even in the winter time.

Look at it this way, suppose we had a factory making ice and that ice was loaded on a train to take it somewhere. The arctic is the ice factory, the atmosphere is the train and heat provides the fuel to run the train. Of course, it's a lot more complicated than that, but that illustrates the basics of it.

Unfortunately, it's even worse than what I've described here. That extra energy in the atmosphere will also mean more energy to move air masses the rest of the year. It takes a lot of energy to put water vapor in the air. As the planet gets warmer, there is more water vapor in the air which means more snow in the winter and bigger rains the rest of the time.[51] This will result in more severe storms all year, not just in the winter. And, they will be more severe, on average, than they were in the past.

I always wonder about the people that make the same old jokes about global warming when there is record cold weather or a big snowstorm in the winter when its suppose to be cold. But, when there is a really warm day in the winter, again, when it's supposed to be cold, no one says a word. Whenever there is a winter storm I always here someone make some ridiculous joke about global warming.

I wonder, just when are they going to stop chuckling and realize we are in a mess of a situation? Or, at least, make some new and original witticism.

From: Fred.Spencer@email.com
To: Dialogue@email.com
Subj: Re: Winter weather
Date: Saturday, May 11, 2013 5:36:28 PM

You're saying that without global warming we wouldn't be seeing all of this violent weather we've seen the last several years?

From: Aileen.Gulledge@email.com
To: Dialogue@email.com
Subj: Violent weather
Date: Saturday, May 11, 2013 7:11:06 PM

No. We have always had violent weather. But, violent weather requires energy. Without energy in the atmosphere there would not be storms at all. So, violent weather is a measure of how much energy is stored in the atmosphere. An increasing amount of violent weather is an indicator that the amount of energy in the atmosphere is increasing. We would certainly see violent weather, even without global warming. But, there is a definite, measurable increase in the amount of violent weather[52,53] and we would not see that increase, or maybe not as big of an increase, if it wasn't for manmade emissions causing global warming.

Global warming is kind of like steroids for the weather. There will be weather anyway, but the additional heat makes everything more extreme. Heat waves are hotter, more frequent and last longer. The same goes for droughts and floods. Tornadoes and hurricanes become more frequent and more severe. Hail storms, too. And, even winter storms.

From: Fred.Spencer@email.com
To: Dialogue@email.com
Subj: Re: Violent weather
Date: Sunday, May 12, 2013 12:34:14 PM

That seems pretty obvious the way you explain it, even without going into details. You would think most people would accept an explanation like that. That was pretty well done, Aileen. Any comments, Tom?

From: Tom.Lawson@university.edu
To: Dialogue@email.com
Subj: Re: Re: Violent weather
Date: Sunday, May 12, 2013 08:14:31 AM

I was going to make a joke about winter storms but I changed my mind.

From: Aileen.Gulledge@email.com
To: Dialogue@email.com
Subj: Temperature
Date: Sunday, May 12, 2013 1:23:51 PM

Global surface temperature is obviously a key indicator of global warming. If the world is warming then we should be able to measure it. And, we can. In fact, we measure it with ground sensors and with satellite sensors.

Ocean temperature is also a key indicator. If the Earth really is warming then the vast majority of the excess heat will go towards heating the oceans. That's because it takes a lot more energy to heat water than air or ground. Nature will try to make everything the same temperature; so much more energy will go into the oceans to raise the temperature of the water by the same amount as land or air.

The amount of ice in the world is another key indicator. We can measure the extent of polar ice in the Arctic Ocean and big ice sheets and we can also measure the amount of ice in glaciers.

Tracking this over a period of time will indicate if the amount is increasing, decreasing or remaining constant.

Both of these last two will affect the sea level. Arctic sea ice is already in the water and melting the ice cap at the North Pole will not effect the sea level any, although it will have other effects on the climate. But, glacial runoff, especially from Greenland and Antarctica, will add water to the oceans if the amount of ice is decreasing. Also, as water heats up it expands. This means that, if the world is warming, we should see the sea level rise as a result. We have good data on this going back many years and we can examine that data to see if the sea level is rising, falling or remaining the same.

From: Fred.Spencer@email.com
To: Dialogue@email.com
Subj: CO_2?
Date: Sunday, May 12, 2013 4:22:08 PM

You didn't mention CO_2 levels. I know this is something that is monitored as an indicator. Why did you leave that off the list?

From: Aileen.Gulledge@email.com
To: Dialogue@email.com
Subj: Re: CO_2?
Date: Monday, May 13, 2013 9:39:54 AM

You're right. CO_2 levels are monitored as indicators. However, we are only interested in determining if the Earth is warming right now. We will be discussing the greenhouse effect later on and I thought it would be better to hold off on the CO_2 levels until then. I think CO_2 levels are an indicator only if you prove the greenhouse effect is causing global warming and we aren't there yet.

From: Fred.Spencer@email.com
To: Dialogue@email.com
Subj: Re: Re: CO_2?
Date: Monday, May 13, 2013 2:53:02 PM

OK, as long as we come back to it later.

From: Aileen.Gulledge@email.com
To: Dialogue@email.com
Subj: Data
Date: Monday, May 13, 2013 7:18:30 PM

We will, there is a lot to say about it. But, let me discuss these indicators one at a time. And, Tom, you are right. The amount of data available is pretty overwhelming. If anything, there might be too much. I think a lot of people would be put off by that. Some people would look at the amount of data and wouldn't even know where to start. I'm going to try to stick to one topic at a time in order to make sense of it all.

From: Tom.Lawson@university.edu
To: Dialogue@email.com
Subj: Re: Data
Date: Monday, May 13, 2013 8:42:00 PM

The amount of data that is available is partly because of public demand. People rightly question this issue and want to see for themselves if there is anything to it.

From: Fred.Spencer@email.com
To: Dialogue@email.com
Subj: Re: Re: Data
Date: Tuesday, May 14, 2013 2:59:53 AM

But, once the data and evidence are made available the deniers have to take advantage of it. You can't demand the data and then simply ignore it.

From: Tom.Lawson@university.edu
To: Dialogue@email.com
Subj: False Arguments
Date: Tuesday, May 14, 2013 10:00:25 AM

What were you doing up at 2:59 this morning?

It is true they ignore most of the data, but I'll show you how they also selectively use the data to their advantage. Selectively picking data to support your conclusion is called 'cherry picking' and is a form of false argument.

From: Aileen.Gulledge@email.com
To: Dialogue@email.com
Subj: Global surface temperature data
Date: Friday, May 17, 2013 5:22:09 PM

Let's discuss the global surface temperature first because that makes the most sense. If there is global warming, the temperature has to be rising. But, temperatures at specific locations can either rise or fall or even remain constant over a period of time. That is not an indicator of what is happening on a global scale. We need to get data about temperatures all over the world at the same time then compare that to measurements over time.

We can do this several ways. The most obvious is to have a bunch of weather stations around the world, record the data and share it with everyone. The first problem with this is making sure all of the instruments make measurements that are consistent with each other. If two instruments are sitting next to each other in a lab it's easy to see if they are working the same, but how do you do that when they are thousands of miles apart? There needs to be confidence that different instruments are working equally well. Also, there are vast areas of the planet where we can't have a lot of weather stations, out in the oceans, for instance.

A second way to make global measurements is to use instruments on satellites and this way can measure the world-wide temperature pretty quickly. But, the measurements are not simultaneous over the entire planet; it takes time for the satellite to see the whole planet. Also, satellites don't work forever and have to be replaced. You then have to make sure the readings of

67

follow-on instruments are the same as the earlier ones.

Finally, we can use indicators in the historical record to deduce what the temperature was in the past. This is a difficult task that takes time to do. It can also be controversial.

Because of these kinds of problems scientist work on not only resolving the issues, like the ones I mentioned, but also finding different ways to get the same measurements. If we can use multiple methods to measure something we can compare the results to see if they are in agreement.

From: Fred.Spencer@email.com
To: Dialogue@email.com
Subj: Re: Global surface temperature data
Date: Friday, May 17, 2013 8:12:07 PM

Wow! You almost sound like Tom!

BTW Tom, I had some overseas business to take care of.

From: Aileen.Gulledge@email.com
To: Dialogue@email.com
Subj: Re: Re: Global surface temperature data
Date: Friday, May 17, 2013 8:45:34 PM

Reading all of this science stuff will do that to you. I have a better understanding of why Tom likes to be so precise in what he says.

From: Tom.Lawson@university.edu
To: Dialogue@email.com
Subj: Accuracy
Date: Friday, May 17, 2013 10:49:15 PM

A small change in the wording can have a big difference in the meaning. What do you have on global surface temperature? With all of those issues I would guess the data could be interpreted many different ways.

From: Aileen.Gulledge@email.com
To: Dialogue@email.com
Subj: Global Surface Temperature
Date: Sunday, May 19, 2013 2:57:19 PM

That would be true if we used only one source. But, when we compare multiple data sources what we find is a very good agreement between them. NASA Goddard Institute for Space Studies, the Met Office Hadley Center (in the U.K.), NOAA National Climate Data Center and the Japanese Meteorological Agency all track the global surface temperature and their results are nearly the same.

Here is a plot showing their data. I found it at a NASA website.[54] There are several things I find very interesting about this graph. First, all four centers show the temperature trending upwards very steadily since the 1970s. Also, I think it is significant the data comes from three nations and four different centers. That means there are different people making the measurements in

69

different ways and they all come up with the same results. There are peaks and valleys in the temperature as it fluctuates from month to month, but all four vary in agreement. To me, that is a great authenticator. Four different agencies using different data and we see they don't just show the same upward trend, they also vary in synch over the years. Here it is:

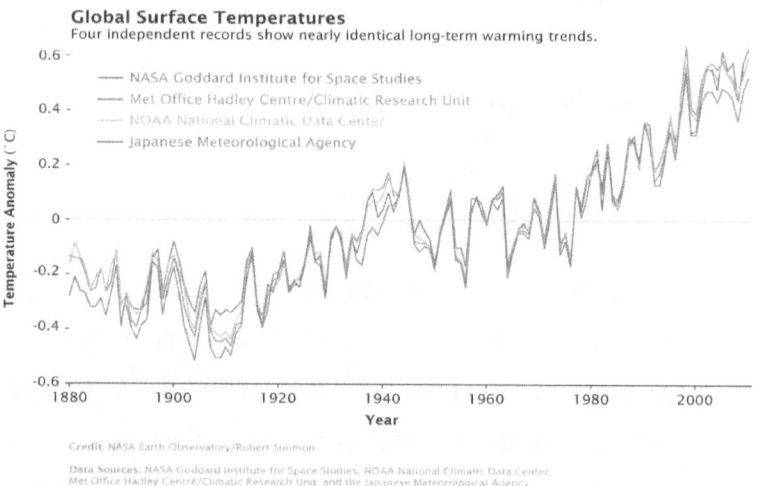

From: Fred.Spencer@email.com
To: Dialogue@email.com
Subj: Re: Global Surface Temperature
Date: Sunday, May 19, 2013 08:14:31 AM

What are we looking at in this plot, Aileen? It is labeled 'Temperature Anomaly' in degrees centigrade and the scale is plus or minus some baseline marked '0'. What does all of that mean?

70

From: Aileen.Gulledge@email.com
To: Dialogue@email.com
Subj: Temperature Trend
Date: Sunday, May 19, 2013 8:24:39 PM

They don't use actual temperatures on this plot. What they do is to average temperature for an area over a period of time, say a month. Then, they compare that to some long-term average. This will show if the monthly average is higher or lower than the long-term average. Each of the centers uses a different long-term average, but that doesn't matter with this method. It only shows the temperature trend. The graph I sent you guys shows the difference from the different baselines for each of the centers.[55] The center line labeled '0' is the long-term average. A point below that would mean the average temperature for that date was cooler than the average. A point above the average line would mean the average temperature for that date was hotter than the average. You can see that the last time there was a temperature cooler than the long-term average was in the late-1970s. It has been hotter than the long-term average ever since.

From: Fred.Spencer@email.com
To: Dialogue@email.com
Subj: Warming trend
Date: Monday, May 20, 2013 5:39:07 AM

I'm not sure I completely understand what you just said, but I get that they are all measuring the same thing with different methods. That is very convincing. You can clearly see the warming trend you mentioned and you can really see that they all get just about the same readings.

From: Aileen.Gulledge@email.com
To: Dialogue@email.com
Subj: Temperature change by hemisphere
Date: Tuesday, May 21, 2013 9:18:39 AM

Here is another one from the NASA Goddard Institute for Space Studies. This one shows the temperature change for the northern and southern hemispheres.[56] You can really see how the average temperature has been going up in the last several decades. The northern hemisphere is warming faster than the southern hemisphere due to the fact there is more land area in the north, but both hemispheres are clearly getting warmer.

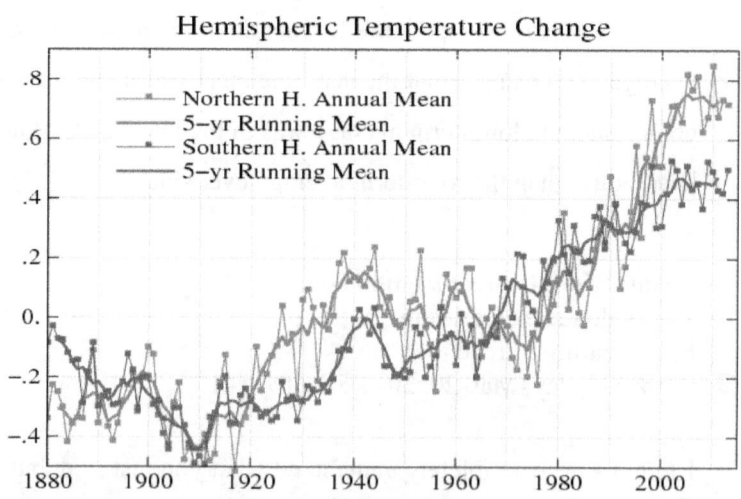

From: Tom.Lawson@university.edu
To: Dialogue@email.com

Subj: Heat island effect
Date: Wednesday, May 22, 2013 10:52:12 AM

My immediate complaint is that you have not addressed the issue of the heat island effect. Thermometers that were originally placed in the countryside are now surrounded by asphalt and concrete, making the local temperature higher and skewing the database. A paper by McKitrick and Michaels claims that the urban heat island affect is responsible for 50% of the measured increase in global temperature.[57]

From: Fred.Spencer@email.com
To: Dialogue@email.com
Subj: Re: Heat island effect
Date: Wednesday, May 22, 2013 5:22:09 PM

That is good point. We all know cities have been growing. Places that were originally in the countryside are now surrounded by concrete. How do you explain that, Aileen?

From: Aileen.Gulledge@email.com
To: Dialogue@email.com
Subj: Re: Re: Heat island effect
Date: Thursday, May 23, 2013 2:32:46 PM

Don't hold back, you guys.

There is little controversy surrounding the urban heat island effect. Everyone agrees that cities get hotter as they get larger and this is separate from climate change. But, studies of rural temperatures compared to urban temperatures have shown that the heat island effect has little impact on the global temperatures.[58] What is interesting is another study which shows McKitrick and Michaels did their analysis wrong. When correct methods were used it was found their work had spurious results.[59] McKitrick and Michaels deny this, but it certainly calls into question their work. When no one else can duplicate your results and independent studies show the opposite result then there has to be a lot of suspicions about their work.

I would like to note a couple of false arguments that are made about this study. Even in their own study, they say the heat island effect is responsible for 50% of global warming. But, they use this as evidence there is no global warming without ever explaining where that other 50% comes from. Also, a common complaint from deniers is that the data is being manipulated, yet here we find the data is manipulated by the deniers in this study, and it is still cited by the deniers as evidence.

There have been some studies of the heat island effect and two things have come to light.[60] First, removing the stations the deniers claim are faulty actually causes the average temperature to go up a little. Quite the opposite of what the deniers are claiming. Also, the areas that are experiencing the greatest amount of temperature increase are the wilderness areas where there are no cities at all, areas such as the Arctic region. If there was any credibility to the heat island complaints neither of these would occur. Take a look at this figure. It shows the amount of temperature change from the average. Areas that are cooler than the long-term average are colored in shades of blue. Areas that are warmer than the long-term average are colored red. You can clearly see the areas that have warmed the most are located away from the cities, areas like Siberia and the Arctic Region.[61]

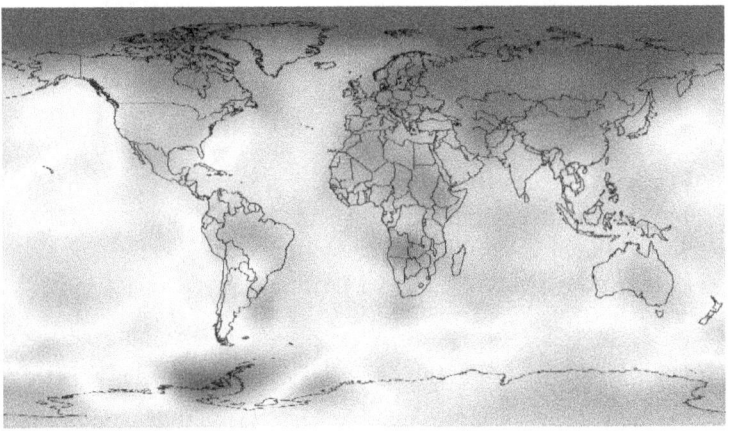

From: Tom.Lawson@university.edu
To: Dialogue@email.com
Subj: Global warming has stopped
Date: Friday, May 24, 2013 11:09:36 AM

OK, but both of these graphs you sent us also show that global warming stopped about 1998. The graphs are basically flat after that and even show a decreasing temperature. Take a look at the paper discussed by the Heartland Institute here: http://www.nipccreport.org/articles/2010/jun/25jun2010a1.html.[62] The authors in this paper, published in a peer-reviewed scientific journal, state, there has been "a decade-long decline (1998-2007) in globally averaged temperatures from the record heat of 1998". And in further describing this phenomenon, they say that U.S. temperatures in 2008 "not only declined from near-record warmth of prior years, but were in fact colder than the official 30-year reference climatology (-0.2°C versus the 1971-2000 mean) and further were the coldest since at least 1996."[63] This really makes it sound as though there is no global warming at all and the data indicates that we are actually cooling.

From: Aileen.Gulledge@email.com
To: Dialogue@email.com
Subj: Global warming has not stopped
Date: Friday, May 24, 2013 11:12:52 PM

That would a wrong conclusion. In fact, 2010 was tied with 2005 for the warmest year on record. And, the first decade of the 21st century was clearly much warmer than any other on record. Not just a little, but a great deal. The data shows the planet is warming.[64] It is true there was a cooling from the record heat of 1998, but this plot shows the heating is still going on. Whether one year is warmer than another is not the issue. The thing that is important is that there is a definite trend over decades of time. We are talking about trends over decades and centuries, not what is happening in an individual year. This is cherry-picking on the part of the deniers, selectively choosing data to support their claims. If you use 1999 as the starting point you get a very different result. That year was much cooler than 1998 and the trend line starting there is much steeper than the one starting in 1998. The deniers don't want to use 1999 because it doesn't support the conclusion they want. This is a good example of how cherry picking is a lie.

I don't think anyone wants to say that the weather and climate are simple natural phenomena. Hopefully, we all agree that they are both very complex with lots of different things factoring into them. Well, if the weather and climate are both complex topics, why would anyone expect climate change to be simple? There are many things involved and this is going to result in some years being warmer and other years being cooler. You cannot pick a handful of years to either refute or support climate change theories. The important thing is the look at the long-term average and that average is pretty clear in showing a long-term warming trend.

The 1980s was the hottest decade ever recorded. But, every year in the 1990s was hotter than the average of the 1980s. Then, every year of the 2000s was hotter than the average of the 1990s. And now, every year of the 2010s has been hotter than the average of the 2000s. Clearly, we are getting hotter and that heating has not stopped. There are many factors involved and it is normal that some years are going to be hotter than others. The important thing is the overall trend, which is rising, and the actual temperatures, which are the highest ever recorded.

Another way to look at the temperature record is to examine the extremes. Extremes are always expected to happen, but as the climate record gets longer certain extremes related to heating are becoming more evident. For example, in the United States, extremes of high temperatures have been occurring at a rate of twice those of cold extremes and that has gotten even worse since June 2010. The average since then has been 2.7 new record highs for every record low. And, in the summer of 2011, that went to a factor of over 8 record highs for every record low.[65] What I am saying is the number of record high temperatures in 2011 was 8 times as high as the number of record lows. By that measure, rising temperatures have not slowed down. If there was no global warming, the number of record highs would be about equal to the number of record lows. That is not the case.

Also, you used a false argument there, Tom. You referred to the temperature in the U.S. in the context of discussing global temperatures. You cannot selectively pick a small geographic area and use it to draw conclusions about the entire planet. We are discussing global warming, not U.S. warming. The evidence in the graph I sent you guys shows the global temperature since 2000 is much higher than 1971-2000 mean you mentioned.

What is interesting about quoting that paper is that you only took selected quotes. I looked up the paper and read what they had to say and their conclusion is much different than what you, and the

Heartland Institute, seem to be indicating.

What the paper said was there was a naturally occurring cooling cycle in the Pacific Ocean that resulted in a cooling of the continents. They believe there will be an increase in global temperatures after this cooling cycle is done. In their concluding remarks, they said it is not likely there will be a prolonged cooling trend in North American temperature in future years.[66]

We see this is turning out to be true. Like I said, 2010 was at least tied for the warmest year on record and may have been the single warmest. 2011 was the ninth warmest. In fact, nine of the ten warmest years on record have all occurred since the year 2000. The only one of the top ten warmest years to occur before this century was 1998.[67]

Let me state here that this is a good example of the importance of actually looking up the paper and seeing what was said. The Heartland Institute took a paper that supported manmade global warming and made it sound as though these scientists were disputing it. They did that by selectively taking what they wanted and ignoring the rest. Nearly all papers are available on line nowadays, so it is pretty easy. I find it difficult to understand a lot of the papers, but I can understand the introduction and conclusion pretty well if I work at it.

What is really interesting is the way The Heartland Institute uses false arguments to make it sound as if they are more credible. They state, "Indeed, there may even be important phenomena operating within this realm of which the entire scientific community is ignorant. And some of those phenomena may well be strong enough to totally compensate for anthropogenic-induced increases in greenhouse gas emissions, so that other natural phenomena end up dictating the ever-changing state of earth's climate, as could well be what has been happening over the last decade or more."[68] Note how they use 'there may' in the first sentence and how the scientific community 'may' be ignorant of them. Then, they use the phrase 'may well' in the second sentence, followed with 'could well be'. In other words, they are trying to take small probabilities and make it sound as if the entire scientific argument is invalid. This is a real misdirection to try and fool the reader into thinking the work by the scientists is invalid. And, really, they never show any evidence that any such possibility even exists.

From: Tom.Lawson@university.edu
To: Dialogue@email.com
Subj: Data gaps
Date: Tuesday, May 28, 2013 8:45:34 AM

It was a nice Memorial Day break, but time to get back to work.

Aileen, how can we be sure of this data? Essentially, we only have data from places where there are lots of people and there are vast areas of the planet where we have little or no data.

Places like the open oceans, the polar caps, even large areas of Africa, South America and Asia.[69] Doesn't this lead to a questionable data set that could lead to the wrong conclusion? Maybe there is no warming at all; it's just something that appears to be there because the data is bad.

From: Aileen.Gulledge@email.com
To: Dialogue@email.com
Subj: Re: Data gaps
Date: Tuesday, May 28, 2013 11:20:34 AM

I had to do a lot of work over the holiday weekend. I want to make sure I get this stuff right for you guys.

Actually, the ocean record is pretty good. Ships at sea make regular reports that include weather and temperature at their location and we have buoys all over the world that make water and air measurements. Satellites provide even more data.

From: Fred.Spencer@email.com
To: Dialogue@email.com
Subj: Re: Re: Data gaps
Date: Tuesday, May 28, 2013 7:42:59 PM

I think Tom's question is still a good one. Even with the ocean measurements, I have to think there are still large areas that don't have regular coverage. If there are problems with the data we would still expect to see all four of the scientific centers come up with the same results.

From: Aileen.Gulledge@email.com
To: Dialogue@email.com
Subj: Data validity
Date: Wednesday, May 29, 2013 2:57:19 PM

That is a complaint deniers make all the time, but it has been shown that different data sets get the same result. Also, the methods have been investigated and found to be valid. A study by Muller and the Berkeley Earth Project examined a different data set using the denier's complaints and obtained the same results.[70] They used over 39,000 stations in their study as opposed to the 7,280 stations used in many studies.[71] The biggest problem with this study is that it was released to the public before being peer-reviewed. This is poor procedure. The deniers point out that peer-review may show errors in the study. However, they made their data, methodologies and analyses programs available to all via their website.

From: Tom.Lawson@university.edu
To: Dialogue@email.com
Subj: Gaps in ocean data?
Date: Wednesday, May 29, 2013 5:22:15 PM

Maybe the existing data base is good. But, we still have the argument that the extent of the data base is poor, especially over oceans, that it is limited and is filled with errors and biases.

From: Aileen.Gulledge@email.com

To: Dialogue@email.com
Subj: No gap in the ocean data
Date: Wednesday, May 29, 2013 7:59:58 AM

But, we are not limited to surface measurements; we also have satellite instruments that can take measurements over the entire planet, including the oceans and the polar caps. You can even see the weather for the Arctic region online.[72] This provides us with yet another method to measure the temperature. If we keep using different methods and keep getting the same results then we have more and more confidence in the data. We can be more confident that what we are seeing is real. The problem would be if we measured the temperature with one method and got a result, then measured it with a different method and got a different result. That isn't what is happening. Different measuring methods are getting the same result.

From: Fred.Spencer@email.com
To: Dialogue@email.com
Subj: Space data accuracy?
Date: Wednesday, May 29, 2013 9:25:47 PM

The question I have to ask is how accurate are these measurements from space? Isn't it possible that there is a design flaw? That would give us equally bad data from all of the sources and fool us into believing that what we are looking at is good.

From: Aileen.Gulledge@email.com

To: Dialogue@email.com
Subj: Space data is accurate
Date: Thursday, May 30, 2013 11:36:07 AM

Obviously, this is the big one and a lot of time is spent verifying the satellite data are accurate and this data is found to be consistent with the surface measurements.[73,74] These data can then be compared to the surface data to make sure both methods are giving the same measurements. Once the data has been verified as accurate it can be used to provide data for any areas where the surface data is missing. All new instruments, on satellites or on the ground, go through a period of testing, calibration and validation before being put into service. This makes us confident in the data and using satellites gives us yet another method showing the global surface temperature is rising.

From: Tom.Lawson@university.edu
To: Dialogue@email.com
Subj: Bad data, bad results
Date: Thursday, May 30, 2013 9:50:08 PM

There are actually quite a lot of questions about the data and the results. The complaint is that when you use bad data you are going to get bad results. Climate Depot is one example of someone that has extensive criticism of the data set.[75] For instance, the early data showed a cooling trend. Also, the data collected since 1998 shows a warming trend, but the upper levels of the atmosphere are cooling.

From: Aileen.Gulledge@email.com

To: Dialogue@email.com
Subj: Good data, good results
Date: Friday, May 31, 2013 11:02:32 AM

The stratospheric cooling has been found to be related to ozone depletion.[76] This is actually very consistent with the global warming models. If anything, this finding helps verify what we understand about global warming.

As for the cooling trend, this has been shown to be incorrect.[77] There was a problem in the data where it was treated as if the satellites passed over the same point on the Earth at the same local time every day, but there was actually some small drift. For instance, it might have been thought the satellite crossed over New York City at 6 o'clock in the evening, but over a period of time this might have changed to 6:10. That example is made up, but it illustrates the real issue. Also, the original researchers made mistakes merging the data from different satellites and it took a number of years before this was found and corrected. When these corrections were made the cooling trend disappeared and the data was found to be very consistent with the historical record.

I think it's important to note that deniers make these claims, but every time the deniers make specific complaints about how the data is bad the scientific community responds appropriately and is able to address the objections. Using the scientific method, there is no doubt about the validity of the data. Any errors that might exist are going to be small and won't change the overall result.

From: Fred.Spencer@email.com
To: Dialogue@email.com
Subj: Re: Good data, good results
Date: Friday, May 31, 2013 2:50:49 PM

Pretty good, Aileen. You make a convincing argument.

From: Aileen.Gulledge@email.com
To: Dialogue@email.com
Subj: Declining Arctic sea ice
Date: Friday, May 31, 2013 8:29:30 PM

But, I'm not done. We don't have to limit ourselves to just the surface temperature, we can check other indicators. If the world is warming we would expect to see the amount of ice in the world decline. In fact, that is exactly what we see.

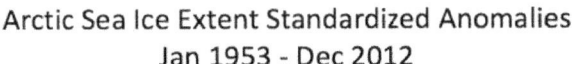

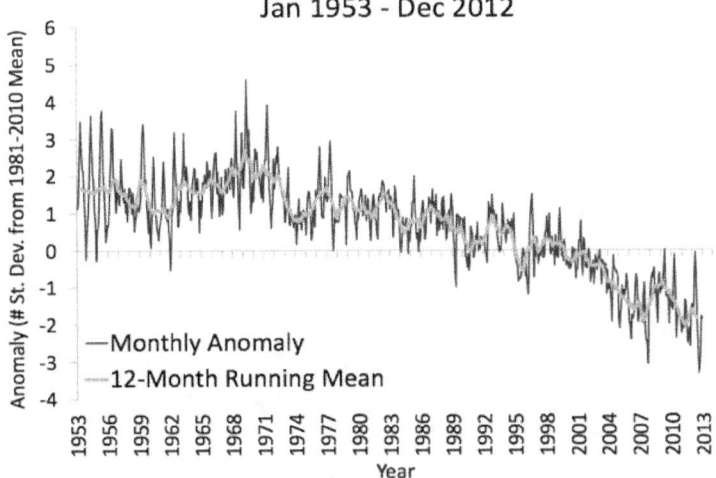

Arctic Sea Ice Extent Standardized Anomalies
Jan 1953 - Dec 2012

—Monthly Anomaly

---12-Month Running Mean

According to this graph from the National Snow and Ice Data Center (NSIDC), the amount of ice covering the Arctic Ocean has been trending downwards since the late 1960s.[78]

The record low was set in 2012.[79] We can see in this graph that the amount of ice cover has decreased by about 40% since 1979 and it continues to decrease. That comes to about 3 million square kilometers of ice that has been lost, more than four times the size of Texas. What isn't shown in this graph is the thickness of the ice which has also decreased dramatically over the same period of time. You can see how much thinner the ice is now than a few years ago in this image below.[80] The lighter the color, the older and thicker the ice. The image on the left shows the ice in 1988 and the image on the right shows the ice in 2013. In March 1988, 26 percent of the Arctic sea ice was thick multiyear ice (4+ years). By 2013 that figure had dropped to only 7 percent.

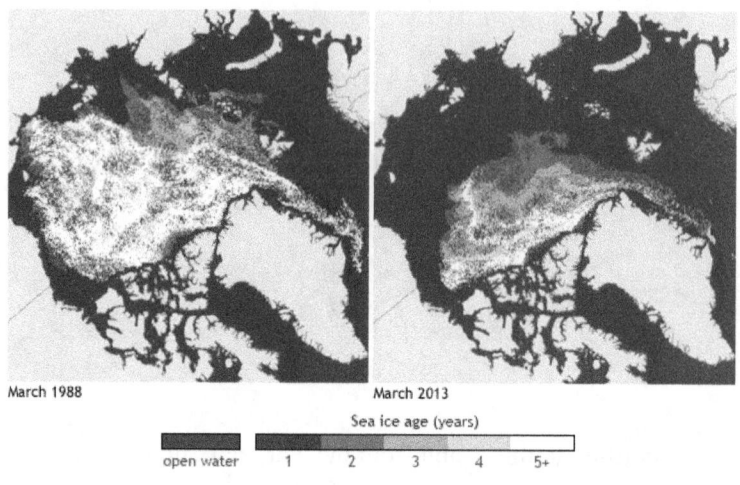

March 1988   March 2013

Sea ice age (years)

open water   1   2   3   4   5+

I came across an interesting story while researching Arctic sea ice about a guy that circumnavigated North and South America in a 27-foot sailboat. Matt Rutherford left Annapolis, MD in June 2011 and sailed all around both continents by taking the Northwest Passage through the Arctic Ocean. It took him about five weeks to successfully navigate from the Atlantic side to the Pacific side.[81] In comparison, the first time anyone successfully navigated the Northwest Passage was the team led by Roald Amundsen. It took them three years from 1903-1906 because the ice was so thick they couldn't get through and then could only make a little progress before being frozen in again when winter set in.[82] This really shows just how much the ice cover has changed in the Arctic Ocean in just the last 100 years.

From: Tom.Lawson@university.edu
To: Dialogue@email.com
Subj: Small data base
Date: Saturday, June 1, 2013 10:55:35 AM

The satellite data has been collected only since the 1970s. So, the data base is small. And now it's even broken because the satellite NASA used to measure sea ice failed in 2009 and won't be replaced until 2016. Plus, we can't rely on the accuracy of these satellite instruments, anyway.

From: Aileen.Gulledge@email.com
To: Dialogue@email.com
Subj: Data base and heat to melt ice
Date: Saturday, June 1, 2013 8:15:08 PM

Again, there is just enough truth there to make it sound plausible, but you don't tell the whole story. Yes, it is true that satellite data began in the 70s and ICESat failed in 2009 and ICESat-2 won't launch until 2016.[83] But, that isn't our only source of data and the database is actually much larger than that. We have data from other sources, such as submarine data going all the way back to the 1950s because the Navy kept records on ice extent and coverage. The gap in satellite coverage is being filled with airborne measurements, known as IceBridge,[84] and the European Space Agency also has CryoSat-2[85] making measurements. Plus, we have teams that go to the Arctic and take actual samples.[86] All of this data is compared to make sure the measurements are correct.

89

Also, measurements from space and samples taken on the ground show the sea ice isn't the only thing melting. The permafrost is thawing.[87] Permafrost is ground that is frozen year round and has been frozen for at least two years. Most of this permafrost is located in the Arctic or high in the mountains. Much of it has been frozen for thousands of years. But, in the last few decades it has been thawing and releasing stored carbon into the atmosphere. In fact, it is estimated that the world's permafrost holds twice as much carbon as is already in the atmosphere.[88] This thawing wouldn't be occurring unless the Arctic was warming. It is even melting in Antarctica where it was expected to be stable.[89]

So, I was curious. Just how much heat would it take to melt an extra 3 million square kilometers of ice with an average thickness of 3 meters. That comes out to be 9 trillion cubic meters of ice. I used metric measurements because it made the math easier.

I used known values for how much energy is needed to melt ice[90] and got 2.75 x 10 to 22nd power joules (2.75 x 10^22 J) of energy. I don't have any context for that, but it sounds like a lot.

From: Tom.Lawson@university.edu
To: Dialogue@email.com
Subj: Magnitude of melting heat
Date: Monday, June 3, 2013 2:51:11 PM

The amount of energy you're talking about would be about 2,700 times the total annual energy production of the United States.

From: Fred.Spencer@email.com
To: Dialogue@email.com
Subj: Melting
Date: Monday, June 3, 2013 5:49:04 PM

Don't you think that is a little advanced for the average person to figure out?

From: Aileen.Gulledge@email.com
To: Dialogue@email.com
Subj: Re: Melting
Date: Monday, June 3, 2013 9:31:41 PM

Why? Just do a search on how much energy it takes to melt a cubic meter of ice. It was pretty easy.

From: Fred.Spencer@email.com
To: Dialogue@email.com
Subj: Amount of ice
Date: Tuesday, June 4, 2013 6:54:39 AM

I really didn't understand the amount of ice that was up there. And, you're not talking about the total amount of ice; this is just the amount that has been lost in the last 30 years. Boy, that's a lot of ice cubes.

From: Aileen.Gulledge@email.com
To: Dialogue@email.com

Subj: Low end estimate
Date: Tuesday, June 4, 2013 10:21:42 AM

That energy figure is only for the amount of energy to melt the ice that has been lost due to the smaller ice extent. What is not included is that the area that is still covered is also thinner. So, we are talking about a significantly larger amount of energy than even that.

Also, that would only be the energy needed to melt the ice when it was at zero degrees and melted into water at zero degrees. Those figures do not include the amount of heat necessary to warm the ice up to zero degrees or any warming of the water after that. And, the ice that doesn't melt is also being heated, even if it doesn't melt. So, this is really a very conservative figure. The total amount of extra energy being stored in the Arctic is actually much more than that amount I came up with.

From: Fred.Spencer@email.com
To: Dialogue@email.com
Subj: Source of heat?
Date: Wednesday, June 5, 2013 5:13:48 PM

One has to ask, where did this energy come from if the world is not warming up? This certainly seems to confirm the previous surface temperature figures. I think you've done a pretty good job proving the world is warming.

From: Tom.Lawson@university.edu
To: Dialogue@email.com
Subj: Maybe not
Date: Wednesday, June 5, 2013 6:25:07 PM

Not necessarily. We now know that there are active volcanoes under the Arctic Ocean. We have to consider that the energy from an erupting volcano could be responsible for the melting ice.

From: Aileen.Gulledge@email.com
To: Dialogue@email.com
Subj: Underwater volcanoes
Date: Saturday, June 8, 2013 10:34:44 AM

OK, I had to go look this up. The area you refer to is known as the Gakkel Ridge. It stretches from northern Greenland across the Arctic Ocean to Siberia and has numerous volcanoes and hydrothermal vents.[91] So, could this volcanic activity provide the necessary heat?

Let's use Mount St. Helens as our reference. The 1980 eruption released an estimated 24 megatons of thermal energy.[92] We need to convert that to joules to keep our figures in the same measurements. It was easy to convert, I just looked up on the web how many joules are in 24 megatons of thermal energy. This comes out to be about 10^{17} joules of energy (1×10^{17} J).[93] When we divide that figure into the amount of heat needed to melt the missing ice we get about 2.75×10^5. That means it would take about 275,000 Mount St. Helens eruptions to account for the extra heat needed to do what we have observed. This calculation does not include any water heating that would occur between the volcano and the surface ice. Even if my calculations are off some, it is safe to say that volcanic activity could not be responsible for the ice melting we have observed.

I would also point out we are getting ahead of ourselves. This is actually a question about natural causes of global warming. Right now we just want to prove the world is warming without discussing the causes.

From: Fred.Spencer@email.com
To: Dialogue@email.com
Subj: Busted
Date: Saturday, June 8, 2013 08:14:31 AM

You are right. I am still very comfortable saying this objection is busted.

From: Tom.Lawson@university.edu
To: Dialogue@email.com
Subj: Total heat is the same
Date: Saturday, June 8, 2013 1:35:00 PM

OK, but there's still more. The extra heat could have come from the other parts of the planet. In other words, the total amount of heat in the climate hasn't changed; it has just been distributed differently. If the wind patterns were to change then more warm winds could blow into the Arctic region and melt the ice. It is not necessary to conclude that there is more heat in the atmosphere. Also, the winds could just be pushing the ice out of the Arctic region and that would account for the decrease in ice cover.

From: Fred.Spencer@email.com
To: Dialogue@email.com
Subj: Local warming
Date: Saturday, June 8, 2013 5:17:09 PM

That's true, Aileen. Just look at the winter of 2011 when the U.S. was having a very warm and mild winter while Europe and Asia where having record cold waves. Why couldn't the Arctic area be warming while other parts of the planet are cooling? There wouldn't be any global warming, just local warming.

From: Aileen.Gulledge@email.com
To: Dialogue@email.com
Subj: Arctic Oscillation
Date: Monday, June 10, 2013 11:43:34 AM

What you are referring to is the Arctic oscillation. Basically, the Arctic region goes back and forth between high pressure and low pressure. This occurs over a period of years. When it has high pressure, it is called the negative phase. Cold air pretty much stays in the Arctic and winters are mild in the lower latitudes. The positive phase is the opposite. There is low pressure over the Arctic and cold air goes down into the populated regions of the planet. Winters in the mid-latitudes, like North America, Europe and Russia, tend to be more severe and warm air goes up into the Arctic.[94] So, what you are saying is that a prolonged positive phase would result in warm air in the Arctic for a number of years and melt the ice up there.

The first problem with this argument is that the Arctic was in a prolong positive phase from the 1970s through the mid-1990s. This is the phase that traps cold air in the Arctic, so the ice extent should have been growing, but was actually decreasing. The phase has alternated between positive and negative since then, but we continue to see decreasing ice amounts.[95]

Also, this argument would contradict the earlier findings that the surface temperature is rising. If warm air is being diverted to the polar region and melting the ice then other regions would have to be getting cooler and that is not supported by the data. And, the iceberg data clearly does not show an increase that would support millions of square kilometers of additional ice. Icebergs in the northern hemisphere mostly come from glaciers in Greenland and a few Canadian arctic islands.[96] If the wind was suddenly blowing more than 3 million square kilometers of additional ice into the North Atlantic or North Pacific this ice would be observed. No such observations have been made. And, keep in mind, there are agencies that monitor this region for ice, such as the International Ice Patrol.[97]

To: Dialogue@email.com
Subj: Not convinced
Date: Monday, June 10, 2013 7:47:24 PM

I'm still not convinced you've completely satisfied the requirements for our first step, yet.

From: Aileen.Gulledge@email.com
To: Dialogue@email.com
Subj: Land ice
Date: Wednesday, June 12, 2013 6:05:04 PM

Fine! I have more evidence. I'm just getting warmed up, so to speak.

Let's not forget there is ice in other places in the world. Mountain snow cover and glaciers are found around the world. If the Arctic is warming because other places are getting cooler we would expect to see the amount of land ice to be increasing, or at least staying steady. But, that isn't what we see.

The amount of land ice is decreasing, also. Glaciers have been monitored since at least 1895 and there is now an extensive data base on many glaciers, although there are still some gaps in the coverage. What has been observed is that there was strong ice loss as early as the 1940s, followed by a moderate ice loss between 1966 and 1985, and accelerating ice losses until the present. The rate of annual ice loss measured between 1996 and

2005 is twice the rate that was observed between 1986 and 1995 and over four times the observed loss rate between 1976 and 1985. There was a period of regional mass gains in the Alps in the late 1970s into the early 1980s and in coastal Scandinavia and New Zealand in the 1990s. These periods are not enough to offset the losses, though and glacial retreat is being observed worldwide, some times dramatically. It is estimated that the world could be deglaciated by the end of the 21st century.[98] A study of 1500 glaciers in Switzerland by Swiss scientists found those glaciers have been shrinking since 1850 and lost 12% of their mass between 1999 and 2009.[99]

Take a look at these before and after images of Pedersen Glacier in Alaska.[100] The one on the left is from 1917 and the one on the right is from 2005. They were taken from almost the exact same spot. You can tell by comparing the backgrounds of the two pictures. This is just one example. Dramatic glacial retreat is being seen all over the world.

Pedersen Glacier

1917   2005

And, data from NASA's GRACE satellite show that the land ice sheets in both Antarctica and Greenland are losing mass. The continent of Antarctica has been losing more than 100 cubic kilometers (24 cubic miles) of ice per year since at least 2002. Greenland is losing 100 billion tons of mass per year and the rate doubled between 1996 and 2005.[101] A study using GRACE data was released in February 2012 and showed that the amount of land ice that melted between 2003 and 2010 was enough to cover the entire U.S. in 1.5 feet of water. About one-fourth of that loss came from glaciers and ice caps outside of Greenland and Antarctica.[102]

If the Arctic is melting because heat is being transferred from other parts of the world then where is the heat that is melting the land ice coming from?

Interestingly, that same study found the ice high in Asian mountain ranges was melting at a much slower rate than had been previously estimated. It is believed this is because the ice is located so high that, while warming, it is still well below freezing.[103] I think this is actually supporting evidence because I would expect just that result if the atmosphere is warming. The lower levels are going to warm to above freezing before the high altitude levels. Suppose, for example, the temperature changed from 0° to 5° Fahrenheit, that is warming, but the ice is still well below freezing temperature, so it wouldn't melt as a result of that warming.

From: Fred.Spencer@email.com
To: Dialogue@email.com
Subj: Ice is melting
Date: Wednesday, June 12, 2013 7:05:09 PM

I like this. Ice is melting in the Arctic, ice is melting in the Antarctic, and ice is melting everywhere in between. That would seem to indicate the world is warming everywhere.

From: Tom.Lawson@university.edu
To: Dialogue@email.com
Subj: Growing glaciers
Date: Wednesday, June 12, 2013 10:10:39 PM

But, studies show that most of the glaciers we have data on are growing.

From: Aileen.Gulledge@email.com
To: Dialogue@email.com
Subj: No, shrinking glaciers
Date: Thursday, June 13, 2013 9:56:56 AM

That is simply not true. The National Snow and Ice Data Center reports that, with few exceptions, glaciers around the world have retreated at unprecedented rates over the last century. Some glaciers and even an ice shelf have completely disappeared. As a point of reference, the 1991 discovery of the 5,000 year-old 'ice man' preserved in a glacier in the Alps shows that particular glacier had retreated to a 5,000-year minimum.[104] He was preserved under the ice and was not being transported down the mountainside. The ice retreated to the point he was exposed again.

From: Fred.Spencer@email.com
To: Dialogue@email.com
Subj: Done?
Date: Thursday, June 12, 2013 2:27:30 PM

Are you done yet?

From: Aileen.Gulledge@email.com
To: Dialogue@email.com
Subj: Ocean warming
Date: Thursday, June 13, 2013 10:15:47 PM

Nope, I did some reading up and what I learned is that if the Earth is warming, the vast majority of the excess heat isn't even expected to go toward heating the atmosphere, but towards warming the oceans. It takes over four times the amount of energy to heat a given mass of ocean water as it takes to heat the same amount of atmosphere. This is called heat capacity and the heat capacity of water is much greater than the heat capacity of air. That means the oceans of the world are the largest storage area for excess heating due to global warming.[105, 106] Therefore, we would expect to see a pretty significant warming trend in the oceans if there is global warming.

As it turns out, the temperature of the world's oceans is something that we measure and record and they are warming.

Because of their very large size, the ocean temperatures are very stable in the short-term and their overall temperature is not easily affected. Any changes would require a long-term, continuous effect. In comparison, temperatures of the land and air can be changed quickly by short-term things that have nothing to do with global warming. We all know how the sidewalk or the beach heats up during the day and cools at night. Well, this creates a great deal of 'noise' in the land and air data that makes it more difficult to see long-term patterns. But, since the oceans are much less sensitive to short-term effects, this level of noise is greatly reduced in the ocean temperatures. This is like the amount of work it takes to move a supertanker as opposed to what it takes to move a little sailboat. The temperature of the oceans would be the supertanker in this scenario and the air and land would be the little sailboats. We can see the land and air temperatures changing by the minute, but the temperature of the oceans changes only a little bit from day to day. Consequently, if we can detect that the oceans are warming over a period of years or decades, this would show there is a long term effect and serve as a positive indicator that the Earth is experiencing a warming trend.

Measurements of the ocean temperatures necessary for making such an evaluation have been recorded since 1955. Naturally, this database has become more extensive and more reliable as technology improves. All of the collected data shows the oceans to be warming over this period of time, with the greatest amount of warming occurring most recently. Some reported data had some problems, but was then corrected. The reliability of the data collected since 1955 has been examined and it has been found that, when data bias and ocean variability is included in the analysis, the data produce a clear warming trend.[107] Comparing several independent data sets reveals they show excellent agreement. So, independent measurements of the world's oceans all show the ocean temperatures are rising and they all agree to the amount of the temperature rise.[108]

This plot comes from NOAA and shows the global ocean temperature. The blue area illustrates the error range. Notice that the error range got much smaller starting in the 1950s.[109]

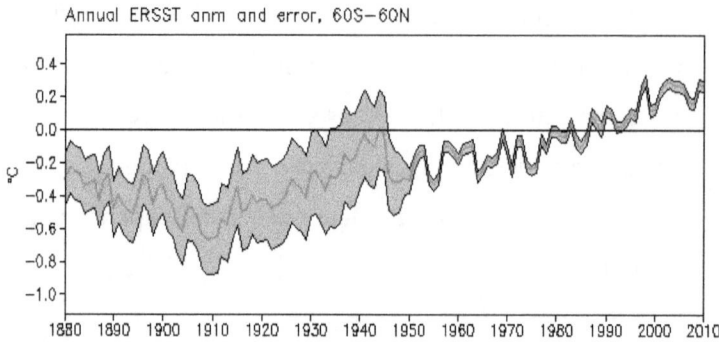

Annual ERSST anm and error, 60S-60N

From: Fred.Spencer@email.com
To: Dialogue@email.com
Subj: Ocean temperature rise
Date: Friday, June 14, 2013 1:35:00 AM

How much of a temperature rise are we talking about?

From: Aileen.Gulledge@email.com
To: Dialogue@email.com
Subj: Ocean temperature rise
Date: Friday, June 14, 2013 10:59:40 AM
A paper on the topic said the authors measured an increase in the heat content of the upper 750 m of the world oceans of about 8.1 x 10^{22} J between 1993 to 2003, followed by a decrease of 3.2 x 10^{22} J between 2003 and 2005.[110] A follow-on paper then corrected this with the finding that the cooling was a mistake due to instrument problems; basically, they didn't calibrate the instruments correctly.[111] After they made the correction to their data they found the cooling period became a warming period and the oceans experienced a warming between 1993 and 2005 that

required an average rate of warming of about 0.33 W/m^2 over the Earth's total surface area.

It is safe to say the world's oceans are definitely warming and this warming trend has been in progress since at least 1955. And, we may safely conclude that the amount of heating that occurred between 1993 and 2003 was about 8.1 x 10^{22} J.

From: Fred.Spencer@email.com
To: Dialogue@email.com
Subj: How much is that?
Date: Friday, June 14, 2013 4:45:35 PM

That sounds like a lot of energy, but is it really? Can you put that into some context that is easier for me to understand?

From: Tom.Lawson@university.edu
To: Dialogue@email.com
Subj: Re: How much is that?
Date: Friday, June 14, 2013 6:54:09 PM

In comparison, 8.1 x 10^{22} J is equal to the entire energy output of about 20 million 1 MT nuclear bombs. A megaton (MT) is a million tons of TNT. In comparison, the Hiroshima bomb was about 18 kilotons (kT). This means each 1 MT bomb is more than 50 times as powerful as the Hiroshima bomb. To get to 8.1 x 10^{22} J in just 10 years, one such 1 MT bomb would have to be detonated every 16 seconds for 10 years straight and require 100% absorption of the released energy.

From: Fred.Spencer@email.com
To: Dialogue@email.com
Subj: OMG!
Date: Friday, June 14, 2013 9:50:09 PM

That seems like an incredible amount of energy. Can that really be true?

From: Aileen.Gulledge@email.com
To: Dialogue@email.com
Subj: Rising sea level
Date: Saturday, June 15, 2013 12:03:54 PM

It is a lot of energy. And, it turns out there is another way to verify this. We can examine the world-wide sea level. As water heats, it expands. If the oceans are really absorbing that much heat we would expect them to expand and we should be able to measure the resulting sea level rise. In fact, that is exactly what has happened.

We have good coastal tide gauge data and this data shows a steady rise in the sea level since the 1870s, rising at an average rate of about 1.70 millimeters per year. But, now we have very accurate satellite data and this shows the sea level rising since the early 1990s at a much higher rate of about 3.27 millimeters per year.[112] The sea level will rise as a result of both runoff from land ice melting and expansion as the water warms up. This provides some very strong supporting evidence that the Earth is warming. Some combination of ocean heating and melting ice is causing the sea level to rise. But, the ice melting is because the Earth is warming. Either way, the rise in sea level shows the Earth is warming up.

From: Tom.Lawson@university.edu
To: Dialogue@email.com
Subj: Scientists change their minds
Date: Saturday, June 15, 2013 4:40:13 PM

This all might be very interesting, but we have already seen how scientists will change their minds. Scientists in the 1970s were predicting we were heading for a new ice age. If scientists were so willing to change their predictions once before, then what's to stop them from doing it again? Clearly, the scientists were wrong back then and they are just as likely to be wrong today.

From: Aileen.Gulledge@email.com
To: Dialogue@email.com
Subj: Scientists didn't predict a new ice age
Date: Sunday, June 16, 2013 3:50:24 PM
This one was interesting; mainly because it turns out to be untrue. It is really remarkable how many of the objections you make that just aren't true, but are repeated over and over. But, let's look at the facts.

A survey of scientific literature between 1965 and 1979 shows that only seven scientific papers were written predicting global cooling.[113] At the same time, there were 44 papers written warning of global warming. There were another 20 papers that were neutral or made no prediction. The principle paper about global cooling that is cited by global warming deniers is a 1971 paper written by a then graduate student at Columbia University named Stephen Schneider, which appeared in the journal *Science*. This paper said that a four-fold rise in atmospheric aerosols could lead to a global cooling. However, we have learned a lot about atmospheric sciences since 1971 and Schneider, now a climatologist at Stanford University, says this new information has made his 1971 paper no longer valid. Schneider says, 'The likelihood of new evidence to overthrow the concept of global warming is small. Warming is virtually certain.'

But, I have heard this claim for years and it just won't die. I wondered where it came from and found an interesting story about it in Scientific American.[114] It turns out this seems to have started with a single, nine paragraph story in Newsweek in 1975. It even became the cover story. It wasn't the scientists making claims about global cooling, it was the press.

So, the evidence shows that, contrary to the claims, the scientific consensus in the 1970s did not predict a new ice age and the majority of the community did not change its mind about the evidence. What happened is the media picked up on this idea and reported it over and over, but the scientific community did not support it.

To: Dialogue@email.com
Subj: Done now?
Date: Sunday, June 16, 2013 9:56:15 PM

Any more?

From: Aileen.Gulledge@email.com
To: Dialogue@email.com
Subj: Daily highs
Date: Monday, June 17, 2013 3:07:54 PM

Yes, there is one more piece of data I've found that I want to talk about and that is daily high temperatures. We discussed it once, but I want to cover it more detail.

If the world is staying at about the same temperature then the number of daily highs should be tapering off. Locally, you could have a record high for the date, but over time this should occur more and more seldom. But, if the planet is warming we should see that there is an increase in the number of daily record highs. At the same time, the number of daily record lows should decrease. This is just the scientific method: make a prediction and put it to the test.

The most reliable way to be sure we are seeing an actual trend is to compare the number of daily record highs with the number of daily record lows. We have already discussed this issue, but we only considered U.S. highs and lows. To support the claim of global warming we need to take a look at the data for the whole planet. And, what we see is that the number of daily highs has been increasing. In the 1950s the ratio between record highs and record lows was almost even, 1.09 record highs per record low. During the 60s and 70s there were actually more record lows than record highs and the ratio was .77 and .78 highs to lows for those two decades. But, since the 1980s this ratio has strongly shifted towards showing more record highs than lows with 1.14 record high for each record low, and continues to increase. Today, there are over 2 record daily highs for every record daily low.[115]

From: Tom.Lawson@university.edu
To: Dialogue@email.com
Subj: Richard Lindzen
Date: Monday, June 17, 2013 6:10:08 PM

But, you haven't even mentioned the arguments of Richard Lindzen. He is the Alfred P. Sloan Professor of Atmospheric Sciences at MIT and is a leading climate change denier. This isn't some crazy guy with a website. And, he says there is no global warming. How do you answer that?

From: Aileen.Gulledge@email.com
To: Dialogue@email.com
Subj: Lindzen not credible
Date: Wednesday, June 19, 2013 8:48:44 AM

Here is an example of Lindzen's lectures:

http://wattsupwiththat.files.wordpress.com/2009/10/cooler_head
s_lindzen-talk-pdf.pdf.[116] One of the things I've learned from
you Tom is to look critically at the 'evidence.' Lindzen's
evidence is really non-existent. Most of his presentation consists
of attacking people advocating climate change. That is not
scientific evidence. He even uses a quote from *The Hunting of
the Snark* by Lewis Carol, the author of *Alice in Wonderland*, to
support his argument. It sounds good, but its not evidence.

What he does bring is some graphs. But, the presentation of
those graphs is made to support his interpretations and he
frequently mixes up weather with climate, two very different
things. The weather of today in Boston is not the same as the
climate in Boston, and certainly not the same as global climate. I
remember a few years ago when we went to see the 4th of July
show and it was so chilly I had to wear a wrap. Surely, no one
would want to take a single evening and use that to make the
point that summers are getting colder! But, that is just what
Lindzen is doing. Scientific reviews of his presentations and
papers find that they are seriously in error to the point it has been
stated his data 'may have been contrived.'[117] That is a really
serious statement in the scientific community. RealClimate.org
has a real good review of Lindzen and his work, showing that he
is not scientifically valid. You can see it at
http://www.realclimate.org/index.php/archives/2007/04/lindzen-
in-newsweek/[118]

What I find most disturbing about Lindzen is a statement he makes that he has not received any funding for his research from the fossil fuel industry and is only funded by the government. The reason this is disturbing is that this is selectively true. The energy industry does not fund his research, but it does fund all of his other activities. He is giving the impression that his work is independent of paying clients, but in fact that is a lie. Why not just come out and state he is receiving funding from the energy industry? Why does he feel it is necessary to mislead people? I just have to believe that someone that would engage in that kind of deception would not hesitate to deceive people about other things. And, that is what the scientific review of his claims indicates.[119] As they say, the lie of omission is still a lie.

From: Tom.Lawson@university.edu
To: Dialogue@email.com
Subj: Lindzen's reputation
Date: Thursday, June 20, 2013 7:18:28 PM

Pretty good. In fact, Lindzen has earned a very bad reputation in the scientific community as someone that manipulates his data and uses clever, non-scientific arguments as 'proof'. I saw a lecture of his that consisted almost entirely of showing why Al Gore was wrong in his movie *An Inconvenient Truth*. Not only were his arguments wrong, but it was irrelevant. That was a movie by a non-scientist. Any errors Gore might have made don't change the science.

He made a number of false arguments in his lecture. For instance, he talked about how China has a larger carbon footprint than the United States. This is true, but what does that have to do with manmade effects on the climate? He talked about how expensive it would be to do anything about greenhouse emissions and how it would ruin the economy. This may or may not be true, but what does that have to do with manmade effects on the climate? He talked about how ethanol is much worse for the environment than gasoline. Again, maybe true, maybe not. But what does that have to do with manmade effects on the climate?[120]

From: Fred.Spencer@email.com
To: Dialogue@email.com
Subj: A convenient truth?
Date: Thursday, June 20, 2013 10:16:17 PM

But, Tom. Isn't it rather convenient for the scientific community to say he is not valid when he disagrees with the scientific community?

From: Tom.Lawson@university.edu
To: Dialogue@email.com
Subj: Scientific validity
Date: Friday, June 21, 2013 1:11:59 PM

He still has to be scientifically valid and his data has to hold up, and he doesn't do either. What the scientific community has to do is allow him to present his arguments and then review them.

It does not have to refrain from criticizing him just because his statements aren't in agreement with most scientists. Lindzen's problem is that scientists can easily debunk him. Of course, his success is that the non-scientific community will believe his clever statements.

But, what about claims the world is actually heading into a new ice age? What do you have to say about that, Aileen? If we are heading into a new ice age it would be hard for us to be warming up at the same time.

From: Aileen.Gulledge@email.com
To: Dialogue@email.com
Subj: New Ice Age?
Date: Sunday, June 23, 2013 12:13:44 PM

This is a great story, Tom. The heart of this claim is an opinion piece in the Australian News.[121] It got a lot of attention because the author is Phil Chapman, a geophysicist and Australia's first astronaut. It makes it sound more credible because of his credentials.

Chapman looked at extreme weather events and the unusually long solar minimum that was going on at the time to conclude we were heading into a new ice age. Unfortunately, there were a lot of things he said that were just flat out wrong.

The first false statement he makes is that the temperature in 2007 was .7C below previous years and that we are now back to the level of 1937. In fact, 2007 was the second warmest year ever recorded at that time.[122]

Mr. Chapman also uses anecdotal evidence that the ice age is coming. He points at snow in Baghdad and the winter in China. As we have already discussed, individual storms will actually get more severe as a result of global warming and the extra energy in the atmosphere. A single data point doesn't prove anything, one way or another.

He also points to the increase in floating sea ice as evidence. Of course, the ice that is being observed is not new ice that is being formed in the winters, it is old ice that is melting off and floating away in the oceans.

But, even with his incorrect data and false arguments, he goes on to make an astounding statement, "It is generally not possible to draw conclusions about climatic trends from events in a single year, so I would normally dismiss this cold snap as transient, pending what happens in the next few years." The reason this is so astounding is that he not only charges ahead and draws conclusions, but he draws some radical conclusions.

He states, "There is no doubt that the next little ice age would be much worse than the previous one and much more harmful than anything warming may do." Actually, there is a great deal of doubt that there will be any next little ice age and there is a great deal of doubt about the magnitude of any such event and the extent of the damage. There is simply no scientific data to support any conclusions of any sort concerning any such event.

He goes on to say, "The next descent into an ice age is inevitable but may not happen for another 1000 years." It is certainly not inevitable that any such event will occur within the next 1000 years, or even within the next million years. His line of reasoning is that the interglacial periods during the last million years have averaged about 10,000 years. This is true, but the periods have varied wildly in extent. Also, there were no ice ages at all for about 165 million years prior to the end of the Cretaceous period. It is a false argument to say that because the interglacial periods of the last million years have averaged 10,000 years that we are due a new ice age because we've gone more than 10,000 years since the end of the last ice age. No data at all suggests that 10,000 years is some magic number.

Continuing in this vein, he states, "...it must be noted that the cooling in 2007 was even faster than in typical glacial transitions. If it continued for 20 years, the temperature would be 14C cooler in 2027. By then, most of the advanced nations would have ceased to exist, vanishing under the ice, and the rest of the world would be faced with a catastrophe beyond

116

imagining. Australia may escape total annihilation but would surely be overrun by millions of refugees. Once the glaciation starts, it will last 1000 centuries, an incomprehensible stretch of time." There is not a single scientifically supported statement in this entire quotation. How in the world can he justify suggesting the average global temperature will drop 14 degrees Celsius in 20 years? Where does he get the idea that this is the end of the world? Where does he get this 1000 centuries figure? He never explains any of this, he is simply throwing it out there.

Then he says, "We cannot really know, but my guess is that the odds are at least 50-50 that we will see significant cooling rather than warming in coming decades." So, what he is saying, even with his incorrect data, false arguments, and unsupported claims, even he cannot be sure that there will be any global cooling. This is an incredible claim to make, considering he wants everyone to change what we are doing. The biggest problem is, there is nothing to support his conclusion.

"It will be difficult for people to face the truth when their reputations, careers, government grants or hopes for social change depend on global warming, but the fate of civilization may be at stake." This is a tried and proven ruse to use in a losing argument. If you cannot refute the evidence, then attack the motives and credibility of those that believe in it. Let's suppose people's reputations, careers and grants depend on global warming. So what? The reverse is not true. Global warming is not dependent on the reputation of climate scientists. Glaciers will melt no matter what scientists have to say about it.

Mr. Chapman is making a false argument by implying the only reason there is a massive amount of data supporting global warming is because we are manufacturing it for personal gain. If this was the case, then Mr. Chapman and other deniers like him would be able to supply scientific evidence to show that the data is mistaken. Why is it they are unable to explain the rising sea temperatures? Why is it they can't explain retreating glaciers all over the world? Why is it they can't explain disappearing sea ice coverage? Why is it they can't explain increasingly shorter winters? Why is it they can't explain how more and more snow fall is being replaced with rain? If this data was being manufactured, it should be easy to demonstrate it.

Finally, all we have to do is look at the record since this piece was published. Sunspots are back and we are warmer than ever. The problem isn't with proving that Chapman was wrong. The problem is that now people will point at this as evidence that scientists are wrong about global warming.

I spent a lot of time on this piece because it is a really good example of false arguments. He just throws stuff out there and doesn't support it with any science, but the science is available for anyone to prove he is wrong.

Wow, Aileen! I'm impressed. I think that was quite a job.

I agree. You did a great job. I'm convinced by your arguments.

Thanks, guys. There is a lot more out there. The toughest part was actually sorting through all of the evidence and cutting it down to size. I could have gone on for a lot longer on just this topic.

Let me finish by summarizing everything.

We saw the Earth surface temperature is rising. This is confirmed by four different groups using different data, with satellite data, and by an independent group. We see that ice is melting worldwide, in each of the polar regions and everywhere in between. We see the oceans are warming and sea levels are rising around the world. And, finally, we see the number of daily record highs is growing and accelerating.

What we have is a large number of independent methods all showing the same result: the Earth is warming.

While it may be possible to dispute the precise numbers, the data clearly indicates that the surface, atmosphere and oceans are warming up and have been doing so for decades. Further, independent data bases show excellent agreement in showing that the amount of heat absorbed by the planet over recent decades is massive in scope.

Hence, the unequivocal answer to our first point is, 'Yes, the Earth is warming.'

From: Fred.Spencer@email.com
To: Dialogue@email.com
Subj: Next
Date: Monday, June 24, 2013 8:23:40 PM

I agree. Now, let's move on to the next question. Is this warming due to the greenhouse effect or is something else causing it?

I never give them hell. I just tell the truth, and they think it is
hell.

Harry S. Truman

Chapter 4
Is the planet warming due to natural causes?

From: Tom.Lawson@university.edu
To: Dialogue@email.com
Subj: You proved step 1
Date: Monday, June 24, 2013 10:14:52 PM

OK. I think you did a pretty good job of proving the world is
getting warmer. I'm sure some deniers might argue there are
alternative explanations for a few of the things you showed, but
there can be only one explanation for everything to be happening
at once. The world must be getting warmer in order to have
everything you showed us to be happening at the same time.

From: Aileen.Gulledge@email.com
To: Dialogue@email.com
Subj: Of course
Date: Tuesday, June 25, 2013 10:38:57 AM

Thanks, Tom. It was tough to sort through all of the evidence. It
wasn't that there was a lack of evidence. The real problem is to
figure out what to talk about and what to leave out. There was so
much scientific evidence that I could still be talking about it.

From: Fred.Spencer@email.com
To: Dialogue@email.com

Subj: Is it something to worry about?
Date: Tuesday, June 25, 2013 4:29:32 PM

I will agree with Tom that you proved the world is warming. But, is this amount of warming significant? Or, is it just some minor fluctuations that we don't have to be concerned with? I don't think you really proved that point yet. We could have a planet that is warming, but it is just a part of a percentage point and nothing that we would need to worry about, or even attribute to a single source.

From: Aileen.Gulledge@email.com
To: Dialogue@email.com
Subj: Amount of heat involved
Date: Thursday, June 27, 2013 9:03:41 AM
That's a fair question and I needed a couple of days to think about how to answer that one. First, let's look at the amount of energy we found it took to melt the ice caps. If you remember, that was 2.75×10^{22} joules of energy. Then, I saw that report showing the upper 2500 feet of the world's oceans had warmed .16 degrees Fahrenheit between 1993 and 2003.[123] I did an estimate of the amount of energy it would take to do this and I came up with 5.1×10^{22} joules.[124] Adding the two gives us 7.85×10^{22} joules. The U.S. generates about 10^{19} joules of energy every year, so it would take the U.S. about 8,000 years to generate the amount of energy absorbed by the oceans and the ice in 10 years. And, that is a low-end estimate.

I think we can say it is a significant amount of energy.

From: Tom.Lawson@university.edu
To: Dialogue@email.com
Subj: Maybe not
Date: Thursday, June 27, 2013 2:31:00 PM

I'm not so sure. There's another way to look at this, if we're going to crunch numbers. The amount of energy from the Sun that reaches Earth every year is about 5.4×10^{24} joules. In ten years, that would be about 5.4×10^{25} joules. 7.85×10^{22} joules would be only .15 percent. That isn't even one-seventh of one percent! That doesn't sound like very much.

From: Fred.Spencer@email.com
To: Dialogue@email.com
Subj: Re: Maybe not
Date: Thursday, June 27, 2013 9:44:54 PM

When you say it like that it doesn't sound like a lot.

From: Aileen.Gulledge@email.com
To: Dialogue@email.com
Subj: No maybes about it
Date: Friday, June 28, 2013 6:41:18 PM

You almost got me with that one, Tom. That is a false argument you are making there. You are trying to say the energy stored isn't large because there is something that is much larger, the energy from the Sun in this case. But, that is irrelevant. The fact that the energy from the Sun is greater than what is being stored doesn't change the fact that energy is being stored. If you want to compare it to something, why don't we compare it to the amount of heat coming from within the planet? The Earth releases about 6×10^{22} joules of energy from its interior[125] over 10 years and that is less than the amount of extra energy stored in the oceans and ice over that same time span.

Viewing it as a percentage of the total amount of energy from the Sun only illustrates how serious this is. These figures show increasing the efficiency of the atmosphere to store heat by just a tiny amount is enough to warm the planet a significant amount.

I also want to point out that if the temperature was stable then we wouldn't be absorbing even an extra .15%. It would be just about zero.

From: Tom.Lawson@university.edu
To: Dialogue@email.com
Subj: False argument
Date: Friday, June 28, 2013 7:35:53 PM

You're right, it is a false argument. And I'm glad you noticed that last point. A lot of people look at the amount of gases we are adding to the atmosphere and say that it isn't important because it is a very small percentage of the total amount of gas in the atmosphere. Well, as you showed, it only takes a little bit to change everything. A good analogy would be drugs, legal or illegal. Even a very small amount compared to the body mass can have a tremendous effect. The same way, even a small amount of extra energy added to the climate can cause lots of changes.

From: Fred.Spencer@email.com
To: Dialogue@email.com
Subj: Significant energy
Date: Friday, June 28, 2013 8:50:06 PM

You went through a lot of math there, Aileen. But, I think you made the point. The planet is warming and the amount of extra energy being stored is significant.

From: Aileen.Gulledge@email.com
To: Dialogue@email.com
Subj: Something is different
Date: Saturday, June 29, 2013 9:27:22 AM

Since we have found that a significant amount of energy has been added to the Earth's environment we can conclude this heating would require some additional cause or source in addition to what was previously in place. In other words, if things had stayed the same the global temperature would have stayed the same. The fact that the planet is getting warmer means something has changed.

From: Fred.Spencer@email.com
To: Dialogue@email.com
Subj: Natural cycle?

Date: Sunday, June 30, 2013 4:43:01 PM

But, is this change due to the greenhouse effect? Or, is it due to some other, unrelated, cause? I've heard the world is just going through a natural warming cycle that has nothing to do with us.

From: Aileen.Gulledge@email.com
To: Dialogue@email.com
Subj: Sources of warming
Date: Monday, July 1, 2013 5:40:21 PM

Remember, we're not addressing the greenhouse effect just yet. We only want to see if there is some source of warming that could explain what is happening. And, don't worry, we will examine the natural cycle claim.

First though, we can divide warming sources into two categories: sources outside the atmosphere and sources inside the atmosphere. Those sources from inside the atmosphere can be further divided into natural sources and manmade sources.

Naturally occurring sources inside the atmosphere would ultimately all require an increased level of warming within the Earth itself, the only true source of naturally occurring internal heating. All other sources of heating would be secondary effects of that one source. When we examine this issue, the first question that has to be answered is whether or not the Earth could experience a sudden increase in internal heating.

Earth's internal heating is caused by radioactive decay of radioactive elements and heat left over from when the planet was created. We can safely assume that these radioactive elements are not uniformly distributed within the Earth's interior and that they move around as internal convection occurs. Therefore, we can conclude there is a possibility that isolated concentrations of radioactive elements are brought together by this convective motion and this would lead to increased levels of radioactive heating. We can't really know the chances of this happening without knowing all the details of what is going on inside the planet.

However, we can be confident that this is not happening. The global warming effects we are witnessing have been widespread and have been occurring on a short time scale, even by human standards and certainly by geologic standards. Any random grouping of isolated radioactive elements inside the planet would take hundreds of thousands to millions of years.

From: Fred.Spencer@email.com
To: Dialogue@email.com
Subj: Internal heat
Date: Tuesday, July 2, 2013 9:56:52 AM

I thought it was real interesting when you said the amount of extra heat we are storing every year is actually greater than the amount of heat being released by the planet itself. That is enough to show that any change in the planet could account for only a little big of the warming, at the very most. But, I also read that only about half of the heat from the Earth's interior comes from radioactive decay. The rest is left over from when the planet was formed.[126] So, even if the amount of heat being generated in the Earth was to change, it couldn't change the amount of heat stored in the environment by very much.

From: Aileen.Gulledge@email.com
To: Dialogue@email.com
Subj: Heating from the planet interior?
Date: Tuesday, July 2, 2013 8:35:17 PM

Good point. And, even if this activity had occurred in the past such that the effects were now being witnessed, this effect would not be something that suddenly appears over the entire planet at once. Measurements of the Earth's surface have not revealed widespread areas of unusual heating. There are isolated hotspots, such as Yellowstone National Park, but they are limited in size. Spots such as these do not produce enough heating to have resulted in the temperature increases we found in step 1, or to even produce anything other than a small fraction of a percentage of this observed heating. Besides, these hotspots have been active for thousands of years and are already included in the global heat budget, so they cannot be contributing to global warming. Additionally, the level of geologic activity on the continents does not show an increase in activity that could possibly account for the observed amount of increased heating.

Further, if global warming was a result of increased heating within the Earth, we would see the oceans warming from the bottom up. The ocean floor is the thinnest crust on the planet and the majority of heating that would escape through the crust would occur there, either through geologic activity such as volcanoes or earthquakes or through direct conduction through the rock. What we see is the opposite; the oceans are warming from the top down. For instance, we can compare modern measurements to measurements taken 135 years ago by the team on HMS Challenger as it made a round-the-world scientific survey trip. Comparing the two shows the temperature of the oceans is about .6 degrees Celsius warmer at the surface, about .4 degrees warmer at 1200 feet and about .1 degrees warmer at 3000 feet.[127] If the oceans were warming from the bottom up then we would find the opposite. The amount of heating would be greater at depth than at the surface. This proves the planet itself is not getting warmer.

From: Fred.Spencer@email.com
To: Dialogue@email.com
Subj: Volcanoes?
Date: Tuesday, July 2, 2013 9:20:28 PM

But, is it possible that the planet is not actually getting warmer, but is still causing global warming? How about volcanoes? Volcanic activity is up and this would cause a lot of heating.

From: Aileen.Gulledge@email.com
To: Dialogue@email.com
Subj: Volcanic activity is stable
Date: Wednesday, July 3, 2013 3:25:54 PM

There is no evidence that volcanic activity is increasing. The number of reported eruptions has increased the last 200 years, but this is due to increased population and better reporting. In other words, the number of eruptions hasn't increased, just the number of reports. In any event, we have had very good scientific measurements of volcanic activity for the last 50 years, and more, and that data shows a very stable amount of volcanic eruptions.[128] So, we cannot blame global warming on increased volcanic activity.

The evidence shows that an increase of internal heating by the Earth cannot possibly be considered to be responsible for more than a small percentage of the observed global warming, at most. Most likely, it is not providing any additional heating over what it has been providing for millions of years.

We're going on a long holiday weekend. I'll be back on Monday. Have a good 4th you guys.

From: Fred.Spencer@email.com
To: Dialogue@email.com
Subj: Slow heating?
Date: Wednesday, July 3, 2013 4:58:41 PM

What if there had been a small increase in internal heating but that heating slowly built up the amount of energy being stored in the environment over a long time?

We can wait for you to get back.

From: Aileen.Gulledge@email.com
To: Dialogue@email.com
Subj: Not a slow event
Date: Monday, July 8, 2013 5:10:55 PM

It's good to be back. I love looking all of this stuff up, but it's hard work and I needed a break. I'm ready to answer some more questions now.

Fred, if that was the case then the heat would have to be trapped in the atmosphere. But, this would require the greenhouse effect or else this additional heating would have been radiated out into space. And, this would prove that the planet is heating up due to the greenhouse effect.

Besides, we have seen that global warming has been happening quickly, not gradually.

From: Tom.Lawson@university.edu
To: Dialogue@email.com
Subj: Human heating
Date: Monday, July 8, 2013 2:26:19 PM

That's a nice, little argument there. But, I'm not satisfied.

Human activity has increased dramatically over the last few centuries, especially since the 1950s, due to an increase in population and industrial activity. In fact, the human population has been increasing almost exactly the same as the increase in global temperatures. There are more cars and power plants and more industry. Landfills release heat and greenhouse gases as everything decays. All of the work to grow crops has been releasing more and more heat. Even if that heat is going out into space we are still making more heat all the time. I think we can account for the entire temperature rise simply by taking into account the amount of heat people release as part of their normal lives. We can explain everything without resorting to global warming theories.

From: Fred.Spencer@email.com
To: Dialogue@email.com
Subj: Re: Human heating
Date: Tuesday, July 9, 2013 9:24:50 PM

That sounds pretty reasonable, Aileen. What do you say?

From: Aileen.Gulledge@email.com
To: Dialogue@email.com
Subj: Amount of human heating
Date: Thursday, July 11, 2013 1:50:43 PM

Nice try, Tom. But, no go. The easy way to test that is to see how much heat humans make and compare that to the amount of heat needed for the observed temperature increases.

According to U.S. government statistics and reported by the Energy Information Administration,[129] world production of energy increased from 3.03 x 10^{20} J in 1980 to 4.85 x 10^{20} J in 2005. This gives an average of approximately 4 x 10^{20} J of energy that human activities release into the environment every year. Taken over 10 years, this comes out to be approximately 4 x 10^{21} J of energy. If you remember, we found in part 1 of our conversation that it would take at least 8.1 x 10^{22} J to heat the oceans as much as has been measured. So, if 100% of global human energy production was retained in the atmosphere for 10 years, it still would amount to less than 5% of the heat needed to heat the oceans. But, that number has a problem because it assumes that all of the heat stayed in the air without escaping into space. This could occur only if we had the greenhouse effect and that would prove step 2. Also, this percentage of heating only reflects the heating of the oceans and does not include any heating of the atmosphere, ice and land masses or heat radiated into space. Further, much of the energy included in the government figures includes energy that would be added to the environment even without man's activities, such as the energy generated through solar power, wind power, and hydroelectric power. All of these activities are directly the result of sunlight and would add heat to the environment anyway. Therefore, we can conclude that, even if the atmosphere retained 100% of the energy released by human activities, it can be responsible for no more than a very small percentage of the observed heating and the cause of observed global warming cannot be blamed on the amount of heat released by human activities.

By the way, this is another argument against the claim Tom made earlier about the amount of heat being involved is insignificant. We can see that all of the heat released by humans combined is just a tiny fraction of the heat required for the global warming.

From: Fred.Spencer@email.com
To: Dialogue@email.com
Subj: Nice work
Date: Friday, July 12, 2013 1:46:54 AM

Wow, Aileen. I'm impressed. That was quite a piece of work there.

From: Aileen.Gulledge@email.com
To: Dialogue@email.com
Subj: Moving on
Date: Friday, July 12, 2013 8:23:12 AM

Thanks. And, all of that information is available to anyone. Tom is right; you don't have to be a scientist anymore to prove this. You just have to be patient enough and willing to do the homework.

But, we still aren't done. We can safely conclude that global warming is not the result of internal heating, either manmade or naturally occurring. This leaves external heating as the only possible alternative source for the observed heating.

From: Tom.Lawson@university.edu
To: Dialogue@email.com
Subj: Hotter Sun
Date: Saturday, July 13, 2013 1:39:45 PM

Well, we know the Sun is getting hotter, so it is only logical that it is leading to a warmer planet.

From: Aileen.Gulledge@email.com
To: Dialogue@email.com
Subj: Cooler Sun
Date: Monday, July 15, 2013 7:59:53 AM

Not true, Tom! The Sun is actually getting slightly cooler.

Dean Pesnell of the Goddard Space Flight Center stated, "Researchers are now seeing the dimmest sun in their records. The change is small, just a fraction of a percent, but significant. Questions about effects on climate are natural if the sun continues to dim."[130]

I saw a number of scientific papers about this.[131] I really didn't understand everything they were saying and, quite frankly, I didn't even want to try. But, I did understand their conclusions. What they found was that solar activity decreased between 1983 and 2003. I loved one paper that said, "Hence, like many authors before us, we conclude there is no credible way that the recent rise in air surface temperature can be attributed to solar effects."[132]

What was really impressed me was they were not using fancy models that you can object to. They used direct measurements.

From: Tom.Lawson@university.edu
To: Dialogue@email.com
Subj: Cosmic rays
Date: Monday, July 15, 2013 8:57:30 PM

You missed one, Aileen. An additional theory concerning global warming and solar activity centers on what is known as the solar/cosmic ray theory. This theory says that cosmic rays create clouds which cool the Earth, but the magnitude of solar activity has increased over the 20th century, providing a better shield against the charged particles that make up cosmic rays. Fewer cosmic rays means less cloud cover. Less cooling cloud cover means the Earth has experienced a resulting warming trend.[133]

From: Aileen.Gulledge@email.com
To: Dialogue@email.com
Subj: Number of cosmic rays are up
Date: Wednesday, July 17, 2013 6:10:41 PM

You are right, I did miss that. There is just so much information out there it is a bit overwhelming. So, I looked it up and this is what I found.

Cosmic rays are not really rays, they are very highly charged particles that travel through space and bombard the planet. Many billions of them hit Earth every second. The solar magnetic field shields us from some, but not all, cosmic rays. The data clearly shows that solar activity has actually decreased over the period of 1983 to 2003 which means the solar magnetic field is not as strong and that leads to an increase in the number of cosmic rays. This resulting increase in cosmic rays has been actually measured by the Ulysses spacecraft that went into orbit around the Sun. So, by this theory, we should be actually seeing more clouds and experiencing global cooling, but we have seen the exact opposite.

As stated in one paper,[134] "Our results show that the observed rapid rise in global mean temperatures seen after 1985 cannot be ascribed to solar variability, whichever of the mechanisms is invoked and no matter how much the solar variation is amplified."

From: Fred.Spencer@email.com
To: Dialogue@email.com
Subj: Variations in our orbit
Date: Thursday, July 18, 2013 11:44:42 AM

But, Aileen, I have read that changes in Earth's orbit could also explain global warming. This is what is believed to have caused the Ice Ages. Why couldn't changes in the Earth's orbit do the opposite and cause global warming?
Happy Birthday, Aileen!

Subj: Happy Bday!
Date: Thursday, July 18, 2013 3:29:18 PM

Yes, Happy Birthday, Aileen

From: Aileen.Gulledge@email.com
To: Dialogue@email.com
Subj: No changes in the orbit
Date: Friday, July 19, 2013 10:21:55 AM

Fair enough. Let's check that out and see if it holds up. This is also part of the 'natural cycle' claim made by some deniers.

Scientists have used ice cores from Antarctica to show that changes in CO_2 levels over the past 800,000 years closely match changes in Earth's orbit.[135] However, no such change in Earth's orbit has been happening over the last 150 years. It is a false argument to say that just because the climate changed naturally in the past over thousands of years, the changes we have observed over the last few decades can therefore be explained as a naturally occurring event. There is no evidence to support that statement.

Thanks for the birthday wishes. The family took good care of me.

From: Tom.Lawson@university.edu
To: Dialogue@email.com

Subj: Sunspots
Date: Saturday, July 20, 2013 3:28:26 PM
What about sunspots and the solar cycle? It is possible to completely describe the temperature history of the entire planet by the sunspot number. I would like to point out that there was a general rise in the number of sunspots over the last few hundred years and this number closely matches the temperature trend. How do you explain this?

From: Aileen.Gulledge@email.com
To: Dialogue@email.com
Subj: Solar cycle effects
Date: Tuesday, July 23, 2013 6:13:08 PM

This is one of those claims that have just enough truth to sound credible. Basically, the Sun is responsible for all global warming. Without the Sun, Earth would be a big frozen, lifeless ball floating around in space. And, in fact, there is good evidence that solar activity at least contributes to variations in the climate.[136] The Maunder Minimum is the period between about 1645 and 1715 when there very few sunspots and is frequently pointed at as evidence of the effectiveness of solar activity towards global warming. This period also coincided with what is dubbed the Little Ice Age, a period when North America and Europe experienced unusually severe winters.

And, there is scientific evidence to show that the Sun's ultraviolet output varies more over a solar cycle than was previously thought. This seems to lend credence to the idea that the solar cycle is responsible for the recently observed global warming, as claimed by some people.

Unfortunately, there is a real flaw in the theory. Several studies have shown that the Sun has been in a slight cooling phase since the 1970s and that trend still continues.[137] But, this was going on at the same time we started to see the unprecedented rise in global temperatures. Never has there been any similar rise in global temperature in the recorded record, going back 800,000 years, as what we are witnessing today. And yet, the Sun actually got cooler during that time span. And, we already discussed the study that showed there is no link between solar activity and the current rise in global temperatures.

Basically, we can say there is credible evidence to believe solar activity and the number of sunspots affected the climate in the past, but the very evidence to support that belief states the solar activity should be leading to a cooling period, not a warming one. So, I think this claim is completely busted. The scary thing to think about is that the warming would be even worse if not for this solar cooling cycle.

That is also part of the 'natural cycle' claim. We can see that, yes, there are natural cycles, but we are actually in a cooling cycle, not a warming one.

Any other objections, Tom?

From: Tom.Lawson@university.edu
To: Dialogue@email.com
Subj: No objections
Date: Tuesday, July 23, 2013 8:12:57 PM

No, I think you did a good job covering the main points.

From: Aileen.Gulledge@email.com
To: Dialogue@email.com
Subj: CO2 levels
Date: Wednesday, July 24, 2013 9:33:05 AM

One last point. As I said, we now have a record of CO_2 levels over the last 800,000 years. At no time have those levels even approached the level we are measuring today. Also, we can compare those levels to the temperature record of the last 800,000 years and we see the two records match. When the CO_2 level goes up, the temperature does also. When the CO_2 level goes down, so does the global temperature. The record is very clear about that. Here is plot from a scientific paper about this[138]:

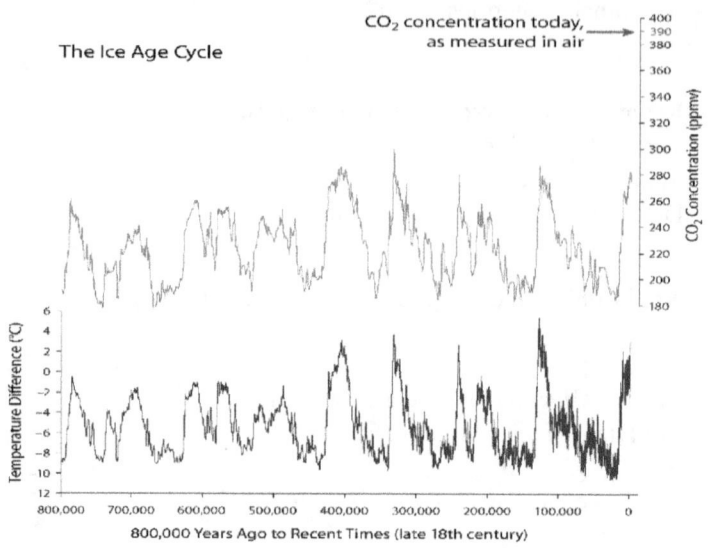

The Ice Age Cycle

CO$_2$ concentration today, as measured in air →

800,000 Years Ago to Recent Times (late 18th century)

Today is on the left-hand side and 800,000 years ago is on the right. The CO$_2$ levels are plotted on the bottom and the temperatures are plotted on top. As a point of reference, our CO$_2$ levels today are about 400 ppm. The highest historical level doesn't even reach 300 ppm. Clearly, humans were not responsible for any of the historical changes. But, that isn't what we are trying to prove at this point. We are just trying to show that the observed global warming is because of the greenhouse effect. I think this last graph really makes the point.

From: Fred.Spencer@email.com
To: Dialogue@email.com
Subj: Conclusion?
Date: Wednesday, July 24, 2013 9:36:17 PM

So, can we conclude that global warming is not a result of an increase in internal heating, heat released by human activity, or the result of external sources. And, we can conclude the greenhouse effect is real?

From: Tom.Lawson@university.edu
To: Dialogue@email.com
Subj: No connection
Date: Thursday, July 25, 2013 11:31:18 AM

But, Aileen, you yourself just proved that those plots don't mean a thing. You showed that there is a link in the past between solar activity and the climate. If all of the climate change in the past was due to this solar activity link that you pointed out, then we can conclude that this plot is just a coincidence and there is no connection between CO_2 and global temperature. If there is no connection in the past, why should there be a connection today?

From: Fred.Spencer@email.com
To: Dialogue@email.com
Subj: Using your own work against you
Date: Thursday, July 25, 2013 3:22:12 PM

Yikes! That is a very good point, Aileen. I'm interested to hear what you have to say. Tom may have just blown this thing out of the water.

From: Aileen.Gulledge@email.com
To: Dialogue@email.com
Subj: Yes, there is a connection

Date: Saturday, July 27, 2013 10:49:30 AM

Oh, come on! It isn't as bad as that.

Tom, you really must listen to a lot of these guys because you keep throwing out all of these false arguments and I know that you are not like that. Once again, this is one of those false arguments that has just enough truth to it to fool people.

The problem with Tom's statement is that it automatically concludes because there is a correlation between solar activity and global temperatures, then all of the climate change is due to solar activity. This is a false argument. In fact, there may be many things linked to climate change, including the level of CO_2 in the atmosphere.

I am sure that no one would say that the weather or the climate is simple. We all know that the National Weather Service does a huge amount of research and computing and still gets the weather wrong more times than not. So, if weather and climate are complicated, why should climate change be simple? The fact is, there are many things involved with climate change and the solar activity is just one, but not the only one.

Just because there is a link between global warming and the solar activity does not in any way refute any link between carbon dioxide and global warming. The reality is, if the solar activity goes up and sends more energy towards Earth, that energy would radiate back into space and global temperature would stay about the same. The only way that the extra energy is trapped and raises global temperature is if there is an increase in the efficiency of the atmosphere to retain heat.

In other words, if the Sun radiates more heat, the planet still needs to retain that extra heat before it can get warmer. So, the greenhouse effect of the atmosphere has to get better or there would be little change in global temperature. The way to do that is by increasing the level of greenhouse gases in the atmosphere. Historically, that extra gas has been CO_2.

And, we do know there is a link between raising temperatures and the level of greenhouse gases in the atmosphere. Raising temperatures release more greenhouse gases stored in the environment, which causes the temperature to rise, which causes more greenhouse gases to be released. For instance, the Arctic tundra is melting and releasing methane.[139] In fact, the amount of CO_2 the oceans can absorb is dependent on the temperature. As the temperature goes up, the amount of CO_2 they can hold goes down.[140] So, increased solar activity will cause the CO_2 levels to go up, which will hold more of the heat in the atmosphere, and the cycle will continue increasing until something changes.

But, in the case of today, solar activity would actually lead to a natural reduction in the amount of CO_2 in the atmosphere. The only reason it is going up is because of us. You cannot try and blame a natural cycle for an unnatural series of events.

From: Fred.Spencer@email.com
To: Dialogue@email.com
Subj: CO2 relationship
Date: Saturday, July 27, 2013 3:48:35 PM

I was little worried there, Aileen. But, I feel good about what you just said. The way I understand it, you are saying the solar cycle in the past would get active at about the same time the CO_2 level went up. And, as the CO_2 went up, it retained more of the solar energy. But, that can't be the explanation for what we are seeing because the solar activity would actually lead to lower CO_2 levels. The fact we are seeing greater CO_2 levels is because of human activity. Is that about right?

From: Aileen.Gulledge@email.com
To: Dialogue@email.com
Subj: Re: CO2 relationship
Date: Saturday, July 27, 2013 4:02:42 PM

Yes. Just because there is a relationship between the solar activity and global temperatures in the past does not mean there isn't also a relationship between CO_2 and global temperatures. There can be both, and there is.

From: Tom.Lawson@university.edu
To: Dialogue@email.com
Subj: Step 2 proven
Date: Sunday, July 28, 2013 8:03:33 PM

Very good, Aileen. I think I am just about tapped out on objections. You did a good job.

From: Aileen.Gulledge@email.com
To: Dialogue@email.com
Subj: Conclusion
Date: Sunday, July 28, 2013 10:29:38 PM

Thanks, Tom. It was a challenge, but certainly one that anyone could do, if they wanted to.

So, let me recap. I eliminated other possible sources of the additional heat leading to global warming. What we have found is that the amount of energy coming in from outside Earth is actually decreasing, while the amount of energy being released within Earth's environment is holding roughly steady. Even the combination of all of these factors cannot account for the additional amount of heat that we see has been added to Earth's environment over the last several decades. And, any argument that these sources are the cause of global warming can be considered ONLY if we assume that the Earth has become more efficient at storing heat. This conclusion would require an increase in the greenhouse effect and would prove point number 2.

In other words, the Earth is not warming because of additional heat input, but because of more efficient retention of the heat that is present.

Since this is the definition of the greenhouse effect, we have shown that the second point is true. We have now proven that global warming is occurring and that this global warming is due to the greenhouse effect.

You do not really understand something unless you can explain it to your grandmother.

Albert Einstein

Chapter 5

Is the planet warming due to manmade emissions?

From: Aileen.Gulledge@email.com
To: Dialogue@email.com
Subj: Keeling curve
Date: Monday, July 29, 2013 9:26:25 AM

I think we cans safely say we have proved the first two points. The Earth is getting warmer and this warming is due to the greenhouse effect. Now, the question we have to prove is that manmade emissions are contributing to this. We have to find out if it is possible the planet is just getting warmer on its own without any help from us.

It is important at this time to point out that the third point in our little exercise does not require the proof that manmade emissions are responsible for all of the observed global warming, or even most. We only need to show that manmade emissions of greenhouse gases cause increases in global temperatures. The exact percentage of warming due to man's emissions is not important to the task at hand. But, I think we'll see that we are responsible for most, if not all, of the observed warming.

What has been observed is that the amount of greenhouse gases in the atmosphere is increasing, and this increase has been going on for as long as we have been measuring the greenhouse gases in the atmosphere at many different locations around the world. These measurements have been made at Mauna Loa in Hawaii since the 1950s and have produced the famous 'sawtooth' plot of CO_2 concentrations known as the Keeling curve, named after R. F. Keeling. Here is an example of the plot from NOAA[141].

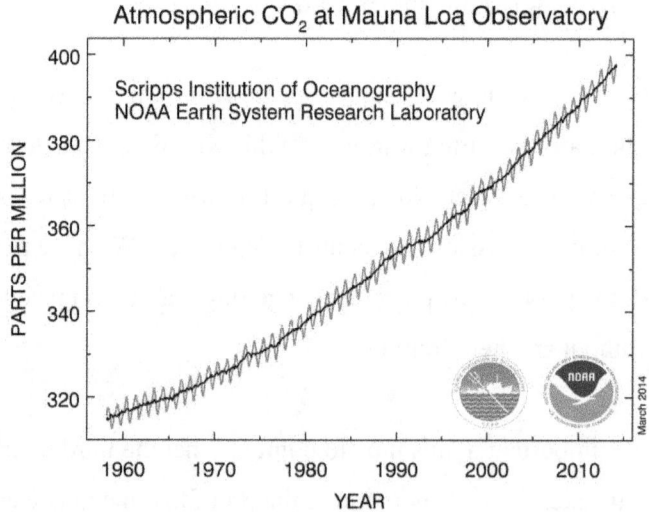

During this same period, man's emissions have also been increasing dramatically. This alone is sufficient to prove point 3 above.

From: Tom.Lawson@university.edu
To: Dialogue@email.com
Subj: Skeptical

Date: Monday, July 29, 2013 5:33:22 PM

All this shows is that the level of CO_2 in the atmosphere is going up. It doesn't prove we are responsible for it. I don't think you can prove to me that this rise in CO_2 is the result of anything we have done.

From: Aileen.Gulledge@email.com
To: Dialogue@email.com
Subj: CO2 fingerprint
Date: Monday, July 29, 2013 7:23:10 PM

Actually, that has been proven. All atoms are not the same. Some atoms are different from atoms of the same element. These are called isotopes. What has been found is that the naturally occurring CO_2 has one kind of isotope while manmade CO_2 has different isotopes. When the increase in gases is examined in detail it is found the extra gas has the isotopes of manmade CO_2.

From: Fred.Spencer@email.com
To: Dialogue@email.com
Subj: Fingerprints?
Date: Wednesday, July 31, 2013 8:55:39 AM

Let me make sure I have this straight. You are saying that molecules have something of a fingerprint based on the atoms that make them up?

From: Aileen.Gulledge@email.com
To: Dialogue@email.com

Subj: Re: Fingerprints
Date: Wednesday, July 31, 2013 1:20:09 PM

Yes, that's right. They are not actual fingerprints, of course. But, they are identifying characteristics.

From: Fred.Spencer@email.com
To: Dialogue@email.com
Subj: Different fingerprints?
Date: Wednesday, July 31, 2013 8:44:37 PM

And, these 'fingerprints' between the naturally made CO_2 and the manmade CO_2 are different?

From: Aileen.Gulledge@email.com
To: Dialogue@email.com
Subj: Re: Different fingerprints?
Date: Wednesday, July 31, 2013 9:02:40 PM

Yes.

From: Fred.Spencer@email.com
To: Dialogue@email.com
Subj: I think I'm getting it
Date: Thursday, August 1, 2013 12:15:02 AM

And, when they look at the CO_2 in the air we find out the extra amount has the fingerprint of the manmade CO_2?

From: Aileen.Gulledge@email.com
To: Dialogue@email.com
Subj: You got it

Date: Thursday, August 1, 2013 11:18:51 AM

Yes again.

From: Fred.Spencer@email.com
To: Dialogue@email.com
Subj: Explain 'fingerprint'
Date: Thursday, August 1, 2013 4:20:46 PM

Briefly, what is this 'fingerprint' that you mention?

From: Aileen.Gulledge@email.com
To: Dialogue@email.com
Subj: Isotopes
Date: Friday, August 2, 2013 12:45:51 PM

Atoms of an element all have the same number of protons in the nucleus, but they can have different numbers of neutrons. They are still the same element, but they have a different mass and have different chemical properties. These are called isotopes. Naturally occurring processes put one kind of isotope in the air while burning fossil fuels puts a different one. By tracking the amount of the isotopes we can identify if the source of the CO_2 is natural or manmade.

From: Fred.Spencer@email.com
To: Dialogue@email.com
Subj: Pretty conclusive
Date: Friday, August 2, 2013 1:37:57 PM

The way you explain it, there is no other conclusion except that manmade emissions are responsible for increasing the level of CO_2 in the atmosphere.

From: Aileen.Gulledge@email.com
To: Dialogue@email.com
Subj: Re: Pretty conclusive
Date: Friday, August 2, 2013 4:10:59 PM

That is the only conclusion. Tom?

From: Tom.Lawson@university.edu
To: Dialogue@email.com
Subj: Conclusive, but who cares?
Date: Saturday, August 3, 2013 7:36:10 AM

You are right. That has to be the conclusion. But, why do we care about the manmade gases? The total amount of manmade gases is only a small fraction of the total gases emitted every year. Natural sources emit far more than all of the gases emitted by human activity.

From: Aileen.Gulledge@email.com
To: Dialogue@email.com
Subj: Natural vs Manmade sources
Date: Monday, August 5, 2013 11:51:22 AM

That is a misdirection to try and confuse the issue. The question isn't if the majority of emissions come from us, the question is, are our emissions changing the environment. The amount coming from natural sources is irrelevant. These naturally occurring sources are balanced by naturally occurring places that absorb those gases. As an analogy, suppose we have a large, old-fashioned balance scale, like the one held by the statue of Justice. Imagine there is a great, but equal, amount of weight on either side. Since the two weights are equal, the scale is balanced. In this scenario, a small weight could be added to one side or the other and tip the scales. This is the situation with greenhouse gases.

The production and absorption in nature is equal. However, the naturally occurring places that absorb CO_2 take in some of the manmade gases, but they can't take all of them. To make things worse, as temperatures go up, new naturally occurring sources begin to appear. Places in the Arctic region where greenhouse gases are frozen, for instance, will become sources of greenhouse gases as they warm up. Other places that absorb greenhouse gases, such as the oceans, will absorb less and less. In this way, the significance of the effects of manmade emissions goes far beyond the percentage of manmade emissions in the total mix. Really, this is one of the greatest threats of global warming.

From: Tom.Lawson@university.edu
To: Dialogue@email.com
Subj: Natural extreme events
Date: Monday, August 5, 2013 4:12:13 PM

Still, if everything in nature is balanced then an unusual event will throw nature out of balance. One large volcanic eruption will produce more carbon dioxide than all of human activity will produce in years, maybe decades, or even centuries.

From: Aileen.Gulledge@email.com
To: Dialogue@email.com
Subj: Volcanic eruptions
Date: Tuesday, August 6, 2013 9:40:08 PM

What? Years, decades, centuries? There is a big difference there, Tom. Can't you make up your mind? It doesn't matter, anyway because the claim isn't true. Just take a look at the Keeling curve and you can see this claim isn't true. If a volcanic eruption put that much carbon dioxide in the air in a just a short time then it would show up in the curve as a spike. But, there are no spikes like that anywhere in the plot. And, there have been some big volcanic eruptions since 1950. For instance, Mount Pinatubo erupted in 1991 and was one of the largest volcanic eruptions ever recorded. But, there isn't any spike in the CO_2 plot. The truth is, most carbon dioxide emitted by volcanoes comes out of the sides and they do it continuously, not all at once[142].

By the way, I noticed it got up to 108 degrees today. It was so hot you could hear the deniers melting.

From: Tom.Lawson@university.edu
To: Dialogue@email.com
Subj: Just a little
Date: Tuesday, August 6, 2013 10:18:48 PM

Is that what I stepped in walking across the parking lot? I was wondering.

But still, manmade emissions are just a small part of the total amount emitted. Surely, you don't expect us to believe such a small amount can have such a big effect.

From: Aileen.Gulledge@email.com
To: Dialogue@email.com
Subj: Not just a little
Date: Wednesday, August 7, 2013 9:34:26 AM

Yes, I do and don't call me Shirley.

The truth is, human emissions aren't that small. For instance, it has been found that the annual CO_2 output of volcanoes is about 145-255 million tons. The annual output by humans is now about 36 billion tons. In other words, humans are emitting almost 150 times as much carbon dioxide as all of the volcanoes in the world combined.[143] Including all geologic sources, humans still produce at least 100 times as much CO_2.[144] The annual human rate of CO_2 production is the equivalent of 17,000 additional Kilauea volcanoes.[145,146,147] And, not only are we doing it every year, but the amount we are producing increases every year.

Studies have shown that approximately 45% of all manmade greenhouse gases are absorbed every year[148]. That is almost half. Using the 36 billion ton annual output I mentioned before, this would result in approximately 19 or 20 billion tons of CO_2 being added to the atmosphere every year by manmade sources. At that rate, we would increase the amount of CO_2 in the atmosphere by nearly 200 billion tons every ten years. It would take all volcanic sources more than 700 years to emit that much CO_2.

So, you can't say that our emissions are just a small amount. Really, it is pretty significant.

From: Tom.Lawson@university.edu
To: Dialogue@email.com
Subj: Might be less
Date: Wednesday, August 7, 2013 5:44:56 PM

But, there is always some amount of error in those kinds of figures. It is possible that the amount of human emissions is a smaller percentage that you make them out to be.

From: Aileen.Gulledge@email.com
To: Dialogue@email.com
Subj: Might be more
Date: Wednesday, August 7, 2013 7:22:32 PM

That is another misdirection, Tom. By your argument, it is also equally possible that manmade emissions are an even larger percentage.

But, even if the levels of manmade versus naturally occurring emissions are not exactly correct, the point has been proven that manmade emissions are significant.

From: Tom.Lawson@university.edu
To: Dialogue@email.com
Subj: So what?
Date: Wednesday, August 7, 2013 9:30:12 PM

All you have done is to prove manmade emissions are significant. You haven't proven they are responsible for global warming.

From: Fred.Spencer@email.com
To: Dialogue@email.com
Subj: Natural cycles
Date: Thursday, August 8, 2013 6:50:50 AM

It sounds good Aileen, I am convinced that human emissions have increased the level of greenhouse gases in our atmosphere. But, I don't think you have shown that this is enough to be responsible for global warming. There is one thing I've heard of that you haven't talked about – the medieval warm period. Doesn't the warming during that period prove that any warming we are going through today could be nothing more than a naturally occurring warm period? It's happened before, why not again. In fact, I believe it has happened many times. Isn't that right?

From: Tom.Lawson@university.edu
To: Dialogue@email.com
Subj: Warm period

Date: Thursday, August 8, 2013 11:03:04 AM

You beat me to it, Fred. I've been waiting to spring that one. And, just let me add that temperatures were higher in the Medieval Warm Period than they are today. This shows there is nothing to worry about. It is just something that happens and nature can take care of it.

From: Aileen.Gulledge@email.com
To: Dialogue@email.com
Subj: Medieval Warm Period
Date: Friday, August 9, 2013 9:35:52 AM

I've been ready for you. This is one of those things the deniers pull out all the time, so I've had a chance to study up on it.

To be clear about what we're talking about, the Medieval Warm Period (also known by many other names including Medieval warm epoch, little climatic optimum and Medieval climatic anomaly) was a time between about A.D. 950 and 1250 when the Northern Hemisphere experienced a very moderate period in the climate. This is the period of time when the Vikings settled Greenland, but those settlements failed when the climate grew colder again, although that wasn't the only reason they failed. The MWP, as it is known, was followed by the Little Ice Age when the Northern Hemisphere was much colder than today. Temperatures during the MWP were comparable to the temperatures of the early-20th century, but less than what are measured today.[149] Here is a plot of the temperature for the

Northern Hemisphere for the last 2000 years[150]:

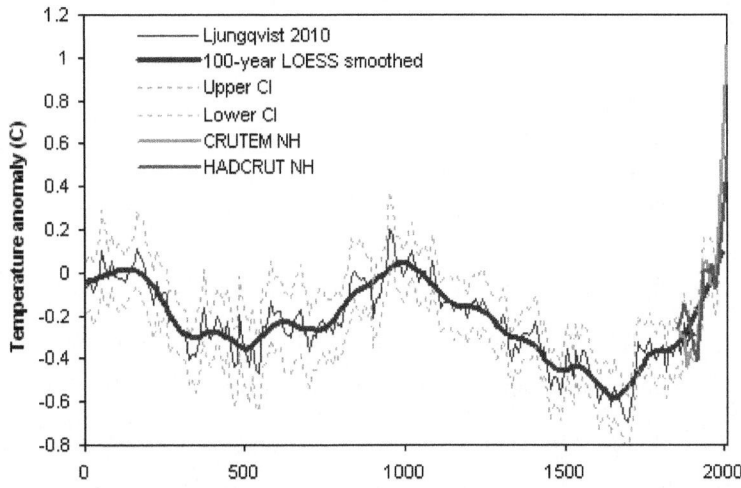

You can see the big hump centered on about 1000 A.D.,
followed by the big decline and then ending with the huge spike
that represents modern day temperatures. You can also see
another bump that went until about 300 A.D. known as the
Roman Warm Period.

And, Tom, I want to make sure I address your comment. The
temperatures of today are much higher than anything that
happened during the MWP.

From: Fred.Spencer@email.com
To: Dialogue@email.com
Subj: Source?

Date: Friday, August 9, 2013 5:53:41 PM

Where do these figures come from? I don't think there were any agencies keeping worldwide temperature figures 2000 years ago.

From: Aileen.Gulledge@email.com
To: Dialogue@email.com
Subj: Temperature proxies
Date: Saturday, August 10, 2013 11:31:55 AM

Obviously, there were no weather stations back then. So, climatologists have to find ways to figure out what the temperature was. They use a number of ways of doing this, including tree rings, ice cores, and mud cores from lake beds and the ocean floors. These are called proxies.[151] They take all of the different data bases they derive this way and compare them to see if they are getting the same results. They have gotten very good at this, but they still keep working at it to make the database even better.[152]

From: Tom.Lawson@university.edu
To: Dialogue@email.com
Subj: Cycles?
Date: Saturday, August 10, 2013 1:28:07 PM

So, when I look at this I see a warming trend about 1000 years ago that lasted for centuries. And, there was another one before that, about 2000 years ago. Two warming periods in a row and we know manmade emissions weren't responsible for either one of them. Why isn't it possible that we are just going through the same thing today, even if it is more extreme?

From: Aileen.Gulledge@email.com
To: Dialogue@email.com
Subj: Yes, there are cycles
Date: Sunday, August 11, 2013 7:45:27 PM

The temperature during the MWP in the Northern Hemisphere was the warmest in the last two thousand years, until the latter part of the 20th century. The average temperature was about what it was in the early-20th century, was less than the period of 1960-1990, and substantially less than what we have seen since 1990. Like you said, Tom, deniers make the claim that this shows warming comes in cycles. It happened before and it will happen again. They want to make the argument that we are merely going through one of these cycles.

Obviously, the deniers are claiming that, since there are cycles, we must be going through a warming cycle. But, there is a real problem with this claim, namely that they are right about the cycles. The climate really does go through cycles, but we are actually in a cooling cycle right now, not a warming one. We have already discussed this very issue and seen the evidence that the current cycle is a cooling one.

From: Fred.Spencer@email.com
To: Dialogue@email.com

Subj: Could be worse
Date: Sunday, August 11, 2013 8:30:29 PM

That is a scary thought because it means things would be even worse without that cooling cycle.

From: Aileen.Gulledge@email.com
To: Dialogue@email.com
Subj: Will be worse
Date: Sunday, August 11, 2013 9:44:36 PM

Eventually, the cycle will come to an end and a natural warm cycle will start. We don't know when that will happen.

From: Tom.Lawson@university.edu
To: Dialogue@email.com
Subj: Rapid warming during the MWP
Date: Monday, August 12, 2013 9:14:33 AM

Still the warming during the MWP is even more rapid than what we have seen in the last 50 years.

From: Aileen.Gulledge@email.com
To: Dialogue@email.com
Subj: MWP is irrelevant
Date: Monday, August 12, 2013 9:00:51 PM

But, that is not correct. We are warming faster than they did then, and the amount of warming is also greater.

Again, you have to read the documents to get the whole story and when I did, I found the deniers always leave out something important.

There is some debate about how extensive the MWP was. It use to be thought that the MWP was limited to only the northern Atlantic and European regions, but the evidence is building that it was actually a global event. Temperature records indicate that the northern Atlantic experienced the most increase in warming, while other places experienced cooler temperatures. But, there is also evidence of warming in the Pacific Ocean and in the tropics, which indicates it was not limited to just one area.[153] I think that just means the event in the Atlantic was so large it had some effect on just about everywhere.

A fatal flaw in the claim by deniers is that we have a good understanding of what happened during the warm period. They want to wave their hand and just dismiss current warming as being the same as during the MWP. But, we know what caused the MWP and those causes do not exist today.

Several studies have shown that solar activity was higher during the time of the MWP and there was a lull in volcanic activity. As we already saw, increased solar activity can cause the temperature to rise. Volcanoes often put large amounts of ash and gases in the atmosphere that lead to cooling. So, there was a combination of increased heating and decreased cooling going on at the same time. This led to atmospheric and oceanic currents that kept the warm period going.[154]

The point is, the Medieval Warm Period does not get us off the hook. We are experiencing global warming today that does not compare to what occurred back then. The rise in the global temperature we are experiencing and the rise in manmade greenhouse gases in the atmosphere are both unprecedented in the historical record. Never have we seen such an increase in greenhouse gases and never have we seen such an explosive rise in temperature.

We have an understanding of what caused the MWP and those conditions do not exist today. So, we can conclude that what happened during the Medieval Warm Period has no bearing on what is going on today. They are not related.

Anything else?

From: Fred.Spencer@email.com
To: Dialogue@email.com
Subj: Point proven?
Date: Monday, August 12, 2013 11:03:59 PM

I'm good. I think you proved the point.

From: Tom.Lawson@university.edu
To: Dialogue@email.com

Subj: Waiting
Date: Tuesday, August 13, 2013 8:50:27 AM

I'm going to hold off for right now.

From: Aileen.Gulledge@email.com
To: Dialogue@email.com
Subj: Conclusion
Date: Tuesday, August 13, 2013 4:52:47 PM

OK, then. We have already shown that global warming exists
and that this warming is due to the greenhouse effect. We have
also shown that manmade emissions are significant enough to
change the naturally occurring balance of greenhouse gases in
the atmosphere. We may, therefore, conclude that manmade
emissions must result in changes in the environment and we call
these changes global warming or climate change.
I think this proves the point we set out to make.

From: Fred.Spencer@email.com
To: Dialogue@email.com
Subj: Satisfied
Date: Tuesday, August 13, 2013 9:29:38 PM

I agree. I think you did a great job and you really showed that
you have to reject science in order to reach any other conclusion.

From: Tom.Lawson@university.edu
To: Dialogue@email.com
Subj: Not satisfied
Date: Wednesday, August 14, 2013 10:27:07 AM

Not so fast. I have some objections. What you have done is to use data and arguments from the people that support the idea of global warming. You have done very little to address the arguments made by the deniers that, they claim, shows manmade global warming is all a bunch of hooey. In particular, the Heartland Institute has made many claims. I would need you to address these claims before I would say you have proven the point.

From: Fred.Spencer@email.com
To: Dialogue@email.com
Subj: Trouble
Date: Wednesday, August 14, 2013 2:10:01 PM

Oh, oh!

From: Tom.Lawson@university.edu
To: Dialogue@email.com
Subj: Yeah, trouble
Date: Wednesday, August 14, 2013 6:55:07 PM

Yeah, oh, oh!

From: Aileen.Gulledge@email.com
To: Dialogue@email.com
Subj: The fun is about to begin
Date: Wednesday, August 14, 2013 9:08:39 PM

No oh, oh! I've been reading those claims and I know they don't hold water. Bring it on, Tom. I'm ready.

To argue with those who have renounced the use and authority of reason is as futile as to administer medicine to the dead – Thomas Paine

Chapter 6

The Claims of the Deniers

From: Tom.Lawson@university.edu
To: Dialogue@email.com
Subj: Deniers
Date: Thursday, August 15, 2013 3:59:56 PM

We'll see just how ready you really are. (How do you put the sound of an evil laugh in an email?) There are lots of claims by the deniers, so this is going to take a while to cover even a fair amount. Let's start with the things that have been in the news lately.

From: Fred.Spencer@email.com
To: Dialogue@email.com
Subj: Re: Deniers
Date: Thursday, August 15, 2013 7:27:07 PM

That sounds reasonable. And, I could make a list like that just based on the news reports.

From: Aileen.Gulledge@email.com
To: Dialogue@email.com
Subj: Re: Re: Deniers
Date: Thursday, August 15, 2013 9:36:01 PM

I've read the news, too. And, since I have been doing my homework I know they are inaccurate. It is amazing how certain headlines have been ignored, while others have been blown out of proportion. Let's do this, Tom. I'm ready.

From: Tom.Lawson@university.edu
To: Dialogue@email.com
Subj: Ice cap recovery
Date: Friday, August 16, 2013 7:07:09 PM

I'll start with an easy one. The Arctic ice cap grew 60% in 2013 and that shows the predictions of an ice-free Arctic are wrong.

From: Fred.Spencer@email.com
To: Dialogue@email.com
Subj: Ice recovery
Date: Friday, August 16, 2013 10:45:52 PM

I have heard this claim a lot and will admit it has raised some questions in my mind. That is a very large recovery in size of the ice cap. A couple of years like that and the ice cap will be back to the same size it was when they started taking the measurements.

From: Aileen.Gulledge@email.com
To: Dialogue@email.com
Subj: Loss of ice
Date: Saturday, August 17, 2013 9:24:54 PM

It would be good if true, Fred. And, I'm glad you started with that one, Tom. And, you're right, this is an easy one. When I first read these reports I admit to being surprised that there was that much increase in the Arctic sea ice. But, I was not at all surprised to see that there was more ice this year than last year.

The reason for the growth is that the amount of sea ice plunged so dramatically last year. It went far below any reasonable expectations. It is not surprising at all that there was a growth in the amount of ice this year. What we have to do is examine both years compared to other years and ask, 'Are we seeing a decrease in ice or an increase?' Look at the plot of ice over the years. This plot comes from the National Snow and Ice Data Center (NSIDC) and is available to anyone[155].

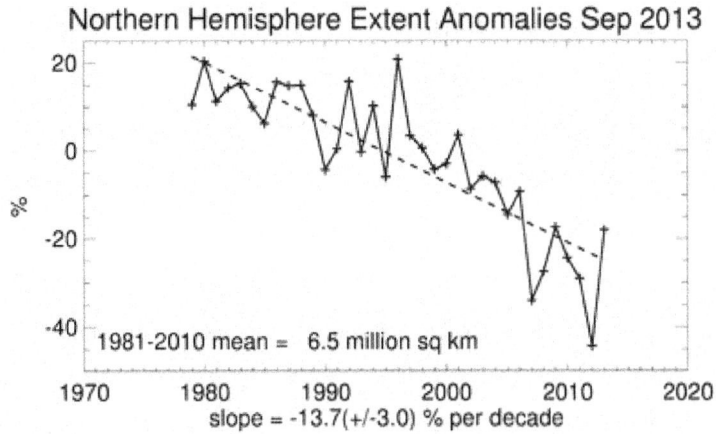

Check out 2012 and 2013, the last two points on the graph and compare them to the straight line. The line is the trend line showing what is happening to the Arctic sea ice on average. You can see that 2012 was way below the trend line. Even alarmingly low. But, we saw something similar in 2007. Notice how the amount of sea ice in 2007 made a big drop and was also alarmingly low. In fact, the only time it has been lower than 2007 was in 2012. But, the amount of ice increased in 2008 and 2009. You could have made the claim that sea ice predictions were wrong and we had two years of increased ice, but that would not be right. Notice that the increase in those two years merely took the amount of ice back to the trend line before it continued to decrease in the years after. This is the same thing with 2013. The amount of ice in 2012 was so low we would expect it to increase to a level closer to the trend line, and that is just what we have seen. In fact, if it hadn't increased scientist would have really been concerned. Granted, a 60% increase is dramatic, but remember how small it was after 2012. A big percentage of a very small amount is still small. Let me ask you, which would you rather have, 60% of one dollar or 60% of a hundred dollars? They are both 60%, but clearly one is different than the other. To say the amount of ice increased by 60% of the 2012 amount is not the same as saying it increased by 60% of the 1995 amount, which is the highest peak on the chart.

I can just as easily say that the amount of ice has decreased by over 30% since that peak in 1995. That would also be wrong. You have to look at the overall picture and not cherry pick your data points in an attempt to make your case.

Here are some facts to consider. In 1980, 62% of the sea ice was thick, multiyear ice. The remaining 38% was thin, first-year ice. By 2012 the amount of sea ice was less than half what it was in 1980 and 58% of it was first-year ice. Another way to measure the amount of ice is to look at the maximum amount of ice before it starts melting in the spring. The amount of ice at maximum is also declining, although at a slower rate. Only 30% of the ice in March 2013 was more than one year old.[156]

The important thing to see is that the trend in the amount of ice is decreasing. Even with the big surge in 2013, the amount of ice is still one of the lowest on record. The deniers want to make a big deal about the rebound but don't want to tell the whole story. The amount of sea ice in 2012 was about 6,000 cubic kilometers. There was a big rebound in 2013 and that figure went up to about 9,000 cubic kilometers. That is about a 50% rebound. But, in the early 1980s there was about 20,000 cubic kilometers of sea ice. So, even with a very large rebound, the amount of ice has dropped a frightening amount in the last 30 years.[157]

It would be wrong for scientists to use 2012 as the basis for their claims, but it is also wrong for the deniers to use 2013 rebound. You have to look at the overall picture and we can see that the overall picture clearly shows decreasing Arctic sea ice.

There is a website, the Polar Portal, sponsored by several Danish institutes, that provides very current data on Arctic sea ice and the Greenland ice sheet. This makes the data available to anyone that wants to look at it. It's at http://www.polarportal.org.[158]

From: Tom.Lawson@university.edu
To: Dialogue@email.com
Subj: NWP transits
Date: Saturday, August 17, 2013 10:48:22 PM

But, what about the Northwest Passage? The Northwest Passage (NWP) is the name of the sea lanes through the Arctic region connecting the North Atlantic Ocean to the North Pacific Ocean. The history of the northern sea lanes tells us that the Arctic Sea has always been a rather difficult ocean to freeze over from time to time. Did you know that the Northwest Passage was successfully navigated by Roald Amundsen in wooden ships in 1905? Then several times by the Canadian wooden ship, St. Roche and again by a smaller Canadian police vessel the St. Roche II circa 1940. Interestingly, the police ship was dispatched by Canada for the expressed purpose of defending Canada's right to controlling the lucrative NWP sea lane should it remain open in the future, which it did not. The feat was also repeated in a luxury cruise ship out of Norway in 1988. Apparently, the variability and difficulty of freezing an entire ocean of salt water in precisely the same way each Arctic winter has been with us for quite a while.

From: Aileen.Gulledge@email.com

To: Dialogue@email.com
Subj: ???
Date: Sunday, August 18, 2013 12:00:49 PM

I'm surprised in you, Tom. That isn't even an argument.

From: Tom.Lawson@university.edu
To: Dialogue@email.com
Subj: Arctic ice loss is not new or unusual
Date: Sunday, August 18, 2013 6:44:34 PM

I know, but this is an actual statement that I have received from deniers. Apparently, their claim is that if a wooden ship could navigate the NWP more than a 100 years ago this somehow shows that the claims of how ships today are traversing the NWP as irrelevant to the issue of global warming. As I interpret the line of reasoning, they are saying the fact that ships can traverse the NWP today is irrelevant because a wooden sailing ship was able to do it 100 years ago. If a wooden ship could do it, then its not surprising that modern, steel ships can do it. The thinning of Arctic sea ice is just a natural occurrence that happens from time to time.

From: Aileen.Gulledge@email.com
To: Dialogue@email.com
Subj: The Northwest Passage
Date: Wednesday, August 21, 2013 8:53:20 AM

OK, I researched it and, to no surprise, I found it is a clearly false argument from top to bottom. Actually, if we examine the facts we see that their argument actually supports global warming.

It is true, Amundsen was the first to successfully traverse the NWP and he did it in a wooden sailing ship. But, he took three years to accomplish this, spending the winters from 1903 through 1906 locked in ice and unable to advance.[159] This was merely the last of many expeditions to attempt this. Before his expedition, many attempts to find the NWP were made over the years. Historically, finding the NWP has been a goal for centuries. While it is known that the Vikings attempted to navigate the waters, the first recorded modern expedition was led by John Cabot under instructions of Henry VIII in 1497. This expedition was followed, over the years, by well over a dozen expeditions, including ones led by both Hudson and Cook. The actual number may be disputed because some expeditions involved exploring rivers in the belief they would lead to the NWP. But, expeditions into the ice were numerous and were attempted from both the Atlantic side and the Pacific side. By the 1770s, the discovery of the NWP was so strongly desired that a reward of £20,000 was offered to the first person to successfully navigate the passage. There was a two-ship expedition in 1845, led by Sir John Franklin, which resulted in the loss of all hands after three years trapped in the ice. The first successful traversing of the NWP was by the crew led by Robert McClure in the period of 1850-1854, but they did not sail the entire route. They actually crossed part of the route by sled, but they were still awarded £10,000 as a reward. As of 1957, only three ships had successfully navigated the NWP. The St. Roche, that you referred to, was the first to transit from west to east. It took two years to complete the voyage, from 1940 to 1942, being forced to spend the winters locked in sea ice. Even though it was fortified for the ice, the captain and crew frequently thought they were about to be crushed by the sea ice. In 1944, St Roche made the return trip, becoming both the first vessel to go both ways, but also the first vessel to make the transit in one season, taking only 86 days. St Roche was the second ship to complete the voyage.[160] The third successful transit didn't occur until 1957, followed by the next successful transits in 1966 and 1977. In comparison, the sea lanes were free of ice during several recent summers, allowing ships to transit straight through without the aid of an icebreaker. Matt Rutherford circumnavigated North and South America in a 27-foot sailboat by himself from the fall of 2011 to the summer of 2012. It took him only a few weeks to take his sailboat through the Northwest Passage.[161] This is vastly different than the situation experienced by Amundsen and all expeditions prior to the 1980s.

As for the Canadian defense of the NWP, of course Canada is going to defend its territorial waters and I cannot see how this applies to the scientific debate. There are several excellent Nova episodes on the Northwest Passage that I highly recommend.[162]

There are also many good books and even webpages that detail the history of the NWP.

So, what we see is numerous attempts to navigate the NWP over a span of several centuries were all foiled by the overwhelming sea ice and even Amundsen didn't conquer the ice, but found a way to live with it for two winters before he was able to get through. In contrast, ships today are finding the passage to be ice-free. In fact, the St Roche II that you refer to transited the NWP in 2000 and was the first to find the passage ice free.

Any attempt to use these historical facts as evidence that there is no global warming is simply invalid. What the evidence shows is that the entire history of the exploration of the NWP supports the theory of global warming because the amount of ice present has dropped dramatically in recent years.

From: Fred.Spencer@email.com
To: Dialogue@email.com
Subj: Cool!
Date: Wednesday, August 21, 2013 10:54:02 AM

Dang, Aileen! That was pretty cool.

From: Aileen.Gulledge@email.com
To: Dialogue@email.com
Subj: Re: Cool!
Date: Wednesday, August 21, 2013 2:31:20 PM

Did you spend a lot of time thinking that one up?

From: Fred.Spencer@email.com
To: Dialogue@email.com
Subj: Re: Re: Cool!
Date: Wednesday, August 21, 2013 7:25:18 PM

No, not really. Still, that was a good job. You convinced me. I
feel like I understand the situation better.

From: Tom.Lawson@university.edu
To: Dialogue@email.com
Subj: Arctic ice is thinning
Date: Wednesday, August 21, 2013 9:24:57 PM

I agree. It was really cool. Also, a good job. I think you made the
point pretty well. Too bad the news media didn't pick up on
these points the same way you did or we wouldn't need to be
talking about it here. The fact is, there really is no news here.
The increase in sea ice would have to continue for several years
before it would be an issue.

From: Aileen.Gulledge@email.com
To: Dialogue@email.com
Subj: That was easy
Date: Thursday, August 22, 2013 1:29:51 PM

Picture me cracking my knuckles at this point.

From: Tom.Lawson@university.edu
To: Dialogue@email.com
Subj: More coming

Date: Thursday, August 22, 2013 8:45:46 PM

Don't get cocky. They will get tougher than this.

From: Aileen.Gulledge@email.com
To: Dialogue@email.com
Subj: Crazy claims
Date: Thursday, August 22, 2013 8:52:29 PM

I know. And, I think that shows how desperate the deniers have

gotten. They really don't have any science to support their

claims, so they are making claims that are crazier and crazier.

What's next?

From: Tom.Lawson@university.edu
To: Dialogue@email.com
Subj: Global warming has plateaued
Date: Friday, August 23, 2013 2:08:19 PM

Global warming has plateaued and there has been no warming in

the last 15 years. In fact, the average air temperatures at Earth's

surface have risen only about 0.2 degrees Fahrenheit since 1998.[163]

From: Fred.Spencer@email.com
To: Dialogue@email.com
Subj: No more warming?
Date: Friday, August 23, 2013 3:48:20 PM

That would be different than what you told us, Aileen. You said

the Earth has been warming over the last several decades.

From: Aileen.Gulledge@email.com
To: Dialogue@email.com
Subj: Warming continues
Date: Monday, August 26, 2013 9:18:32 AM

It has. This is an example of cherry picking the data. Even with the cherry picking it is still wrong. Take a look at the average temperature plot we looked at before.[164]

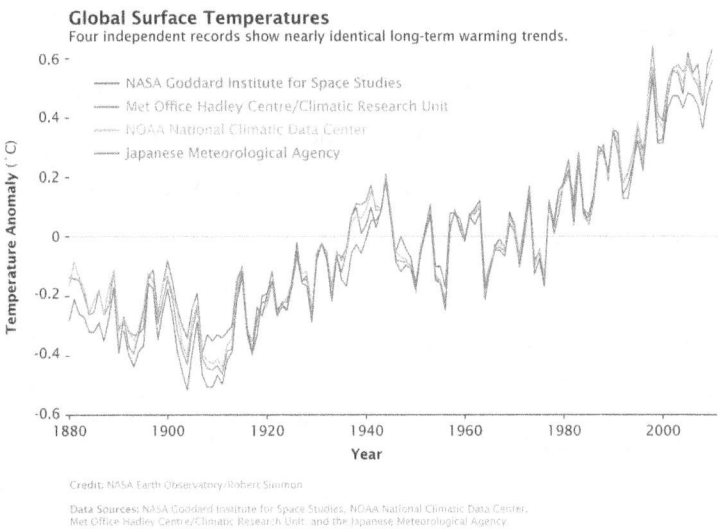

You can see what the deniers are doing. You can see how the temperature for 1998 was way out of whack. It was like 2012 was for the Arctic sea ice – extremely out of line with the rest of the data. What deniers want to do is select this as the starting point to compare global temperatures. This is what is meant by the term 'cherry picking.' They selectively choose data to make the case they want. A valid question would be why they didn't

181

select 1999. We can see that the average temperatures dropped a lot for that year and using that year for the starting point paints a very different picture than if we use 1998 as our starting point. So, why not use 1999 as the starting point?

The answer to the question is that you can't use either year as the starting point. You need to use the average over a period of time. This is the only way to smooth out the data and eliminate odd years that buck the trend, either up or down.

Here are some facts. Nine of the ten hottest years have occurred since 1998, including the three hottest. 1998 is now the fourth hottest year on record. That alone refutes the claim that global warming has stopped.[165]

Here's some more. The 1980s were the hottest decade ever recorded. Every year of the 1990s were hotter than the average of the 1980s, making the 1990s the hottest decade every recorded. Every year of the 2000s was hotter than the average of the 1990s, making the 2000s the hottest decade ever recorded. Depending on the data set used, 2012 was the eighth or ninth hottest year on record since 1850. The globally averaged sea surface temperature was one of the eleven highest on record. The heat content of the upper 2300 feet of the oceans remained at near record levels.[166] The data clearly and consistently shows the planet is continuing to warm. According to a report from the World Meteorological Organization, the first decade of the 21st century was the warmest recorded since modern measurements began around 1850. It also noted the rate of increase in global temperatures between 2000 and 2010 was 'unprecedented.'[167]

The trend is very clear, the world is getting warmer and that includes the last 15 years. For the deniers to say there has been no warming for the last 15 years is simply not true. However, if they want to limit their claims and say atmospheric warming has slowed down, then that is a true statement. The amount the atmosphere has warmed over the last several years has been less than the trend line.

From: Fred.Spencer@email.com
To: Dialogue@email.com
Subj: Missing heat
Date: Monday, August 26, 2013 7:36:54 PM

What does that mean? If global warming has slowed down I would think it means global warming claims weren't what we thought they were. I mean, where is the missing heat?

From: Aileen.Gulledge@email.com
To: Dialogue@email.com
Subj: Oceans are warming
Date: Tuesday, August 27, 2013 11:26:25 AM

That is a valid question, and one scientists have been working on. Scientists are now confident the missing energy is being stored in the ocean, especially the Pacific Ocean. There was a study showing the rate of warming in the deep ocean is greater than the surface of the ocean. [168] A team of scientist have actually found some unprecedented warming of the oceans in recent years that is likely to account for the missing energy. The problem is that we can measure the amount of energy coming in and going out and we know the amount of energy going out of the planet is less than what is coming in. That missing energy must be somewhere. What this team found was that between 2004 and 2008 there was no warming of the ocean between the surface and about 2200 feet. But, the area between 2200 and 6500 feet saw more warming than had ever been seen before. They say this would take about a third of the energy trapped by the greenhouse gases over the last decade.[169]

The key thing, though, is that the evidence shows global warming is still occurring. Claims by the deniers that it has stopped make for good press, but aren't true. And, really, the thing to focus on is the rising level of greenhouse gases in the atmosphere. If you make such a large change in the environment you have to expect to see some changes as a result.

From: Tom.Lawson@university.edu
To: Dialogue@email.com
Subj: New ice age prediction
Date: Tuesday, August 27, 2013 5:59:09 PM

You are just shooting them down. Try this one. One of the arguments by global warming deniers is that scientists in the

1970s were predicting we were heading for a new ice age. Their claim is that if scientists were so willing to change their predictions once before, then what's to stop them from doing it again? Clearly, they say, the scientists were wrong back then and they are just as likely to be wrong today.

From: Aileen.Gulledge@email.com
To: Dialogue@email.com
Subj: No ice age prediction
Date: Wednesday, August 28, 2013 2:31:35 PM

This, of course, is a totally invalid argument and one we have already discussed, but let's discuss it again.

Saying modern climate science is wrong because they made a mistake decades ago would be like saying modern mechanics is invalid because the ancient Greeks made some mistakes. But, as it turns out, it is also just plain not true. A survey of scientific literature between 1965 and 1979 shows that only seven scientific papers were written predicting global cooling. At the same time, there were 44 papers written warning of global warming. There were another 20 papers that were neutral or made no prediction. The principle paper about global cooling that is cited by global warming deniers is a 1971 paper written by a then graduate student at Columbia University named Stephen Schneider, which appeared in the journal Science. This paper said that a four-fold rise in atmospheric aerosols could lead to a downward spiral in global temperature.

However, we have learned a lot about atmospheric sciences since 1971 and Schneider, now a climatologist at Stanford University, says this new information has made his 1971 paper no longer valid. Schneider says, 'The likelihood of new evidence to overthrow the concept of global warming is small. Warming is virtually certain.'

So, the evidence shows that, contrary to the claims, the scientific consensus in the 1970s did not predict a new ice age and the majority of the community did not change its mind about the evidence.[170,171]

And, don't forget we already discussed this and saw that it all goes back to a cover story in Newsweek in 1975.[172] This was something the news media did, not the scientists. I don't know about you guys, but the idea that the news media might get something wrong and then flip-flop their position is not something I find at all surprising.

But, while we are on the topic of changing positions, let's talk about the deniers. If they are going to say climate science is wrong because they got something wrong years ago, then they have to apply the same standard to themselves. The deniers frequently get the facts wrong and then change their stories. And, one of the major sources of funding for the deniers is The Heartland Institute. These are literally the very same people that told us for years that science did not show any danger from smoking cigarettes. In fact, they are even now claiming second hand smoke is harmless. And, we know for a fact that they lied about this because the tobacco industry has come clean on this. So, if the deniers are changing their positions and have actively worked to deceive the public on the scientific evidence then it stands to reason that what they say today is equally invalid.

I mean, that is not my logic, that is the logic of the deniers themselves.

From: Tom.Lawson@university.edu
To: Dialogue@email.com
Subj: 2nd Law of thermodynamics
Date: Thursday, August 29, 2013 4:56:24 PM

Here's another one I get from the deniers all the time: Global warming is a violation of the 2nd law of thermodynamics. Heat cannot flow from a cool body to a warm one without work being done. Therefore, the atmosphere cannot warm the ground.

From: Aileen.Gulledge@email.com
To: Dialogue@email.com
Subj: Really?
Date: Thursday, August 29, 2013 8:47:37 PM

They really want to try and make that argument? This one is really simple. The atmosphere is not warming the ground, and no one is claiming it is. The air is merely trapping the heat coming out of the ground and keeping it from radiating into space. It acts like a blanket on the planet. The thicker the blanket, the more heat gets trapped. Adding greenhouse gases to the atmosphere is like adding more blankets.

From: Tom.Lawson@university.edu
To: Dialogue@email.com
Subj: Models are invalid
Date: Friday, August 30, 2013 10:04:51 AM

Again, I think you covered that one pretty well. You got the right point there. It makes a nice false argument that people will fall for, but it is just a false argument. You really have been doing some homework. Here's the next one. Models are not accurately forecasting what has really occurred, so all predictions of future warming are not valid.

From: Fred.Spencer@email.com
To: Dialogue@email.com
Subj: Need confidence in the models
Date: Friday, August 30, 2013 1:03:48 PM

If we are going to use the results of models as a guide for what we need to do in the future, then this is a very valid issue. Global warming scientists and advocates are saying we need to take all sorts of actions, some of which would cost a lot of money. We need to be sure that those actions are needed before we make the changes. If the claims are based on models then we need to know if the models are valid. What has happened in the past may, or may not, be an indicator of what happens in the future. We really need to clear up this question, Aileen.

From: Aileen.Gulledge@email.com
To: Dialogue@email.com
Subj: Models are valid
Date: Tuesday, September 3, 2013 10:35:41 AM

I think you said it pretty well, Fred. We have to clear up this issue.

The fact is, models are getting quite good. Early models did not give very good results and that shouldn't be a surprise to anyone. Our understanding of the science and the amount of data that was available in the early-1980s was pretty poor. Today, we have a much better understanding of what is going on and much more data for the models to work with.

The claim that models are completely wrong is just not true. Studies show that models agree with actual measurement within 95%. This is very good. The big thing deniers jump on is that models did not predict the slow-down in warming. They argue that this one lapse in the models means they are completely invalid. You can make all sorts of analogies to show how silly this argument is. You could say that doctors couldn't figure out some one disease, therefore all modern medicine is invalid. One bridge failing means all engineering is invalid. One crop failure means all of modern agriculture is invalid. The list goes on.

189

A false argument made concerning climate models is that because they are not completely accurate today, they are therefore as inaccurate as they were 20 years ago. Clearly, we are learning more and more as time progresses. Scientists identify sources of error and address them. Models today still have uncertainties, but the levels of those uncertainties are decreasing and it is reasonable to expect them to decrease in the future. As it is, climate models today are already producing accurate short-term predictions.

Climate modelers also have access to data that did not exist in the past. Satellites such as Aura, Terra and Aqua, and others, provide large amounts of data that were not available until recently. This amount of data from space is being replicated by a corresponding increase in data collected by ground stations. The data available for use in models is now larger, of better quality, and covers a greater time frame than it did 20 or 30 years ago. Models have increased in their reliability as a consequence. Future improvements in data will result in further improvements in the models.

Despite warming over the last 16 years, global surface temperatures have warmed at a slower rate than the previous 16 years. Deniers claim this is evidence that we can't trust the models because they over estimated the amount of surface warming over this period. If they can't get it right today, then how can we trust what they say about 100 years from now?

The important thing to understand about models is that climate models cannot predict the timing and intensity of ocean oscillations like La Niña and El Niño. These are natural cycles that greatly affect global temperature in the short-term by bringing cold water to the surface or sending it down deep. Also, climate models can't predict the timing and duration of volcanic eruptions and industrial pollution, both of which reflect sunlight back into space and reduce the amount of sunlight reaching the surface. When these factors are not included, the models give too much warming. But, when the model input includes these events the models are in very close agreement with actual measurements, including the temperature record of the last 16 years.[173]

From: Tom.Lawson@university.edu
To: Dialogue@email.com
Subj: Scientist are paid to find global warming
Date: Tuesday, September 3, 2013 4:01:30 PM

Maybe so. But, we need to keep in mind that all of this stuff comes from scientists that are getting their funding from sources that want them to prove global warming. The only reason they find evidence of climate change is because they want to keep their jobs.

From: Aileen.Gulledge@email.com
To: Dialogue@email.com
Subj: Arrrghhhh!
Date: Tuesday, September 3, 2013 6:26:16 PM

I know you are just playing a role, Tom, but this one actually makes me a little mad. Saying climate scientist want global warming just to keep their jobs is like saying police want crime so they can keep their jobs, or firemen want houses to burn down so they can keep their jobs. The truth is that climate scientists get paid to study the climate and there will be climate to study with, or without, climate change. In other words, their jobs do not depend on finding evidence of climate change.

In fact, if there were every any evidence that they were acting improperly in their research their careers would be over. Faking or tampering with the data is very unethical and anyone doing that is pretty much done.

From: Tom.Lawson@university.edu
To: Dialogue@email.com
Subj: Won't happen for decades anyway
Date: Wednesday, September 4, 2013 2:37:21 PM
Fine! But, why should we concern ourselves with something that won't happen for decades, if it happens at all? Even the scientists keep talking about the changes in 2050 or 2100, or even later that that. That leaves plenty of time to do something about it. We have other problems we need to deal with right now.

From: Aileen.Gulledge@email.com
To: Dialogue@email.com
Subj: Warming is already happening
Date: Friday, September 6, 2013 7:54:32 AM

Climate scientists frequently refer to some threshold figure in order to describe the effects of global warming. For instance, a common threshold is a 2 degree Celsius rise in temperature over the pre-industrial level, which is cited as the threshold for dangerous changes in the climate. Then, they will talk about some date in the future when we are expected to pass that threshold. I have always had problems with these thresholds because they give the impression that the danger is way down the road and isn't anything we need to worry about today. That is a completely false impression to give. The changes are already happening and just because there are stages that won't occur until after the threshold doesn't mean we can't suffer from some changes right now.

An analogy would be declaring bankruptcy. You don't just wake up one morning and find your finances have so completely fallen apart you need to seek protection with the court. No, things were falling apart for quite some time, getting progressively worse. At first, things weren't bad. Maybe you noticed that you had less and less spare money after paying the bills, but you were still OK. Then, more and more things happened. Maybe you realized you were borrowing money to pay for things that you use to pay cash for. Maybe you were turned down for credit when you applied for it. Maybe your credit card was turned down at the restaurant. At first, it was once a month. Then, it was every week. Finally, it was every day. Then, you found out you didn't have enough money in your paycheck to pay the bills, not to mention buy groceries. How were you going to feed the kids? That is when you sit down and wonder, "How did I get in this mess?"

That is the situation with climate change. Things have started out slowly, but we can already see they are getting worse. A little warming, maybe a few more severe storms than we use to have. Over time, more and more things happen. Mountain snow gets less and less leading to drier months in the summer. More and more crops fail due to heat or are destroyed in more frequent severe storms. You go to the grocery store and are stunned to see the price of food. Droughts occur more often and last longer. Reservoirs that use to be full of water are now full of weeds. When it does rain it comes down so fast that the area is flooded, washing everything away. You notice your electric bill in the summer is going through the roof because you have to run the air conditioner so much. The list goes on, but it is still manageable. We haven't reached the crisis point yet.

But, some day, we will reach that crisis point. When we do, we will realize that we saw it coming for years and didn't do anything about it. Then, we will all sit down and wonder, "How did we get in this mess?"

From: Tom.Lawson@university.edu
To: Dialogue@email.com
Subj: No warming again
Date: Friday, September 6, 2013 3:35:07 PM

Well, how do you explain there hasn't been any global warming
for the last 15 years?

From: Aileen.Gulledge@email.com
To: Dialogue@email.com
Subj: Didn't we just do this one?
Date: Friday, September 6, 2013 5:10:36 PM

What in the world are you talking about? We've already
discussed that one and we have seen that the three hottest
temperatures and nine of the ten hottest years have all occurred
since 2000. How can global warming have plateaued if that is
true?

From: Tom.Lawson@university.edu
To: Dialogue@email.com
Subj: Same song, different verse
Date: Friday, September 6, 2013 8:08:59 PM

The Met Office reported just that. Read this article about it in the
Daily Mail: http://www.dailymail.co.uk/sciencetech/article-
2217286/Global-warming-stopped-16-years-ago-reveals-Met-
Office-report-quietly-released--chart-prove-it.html

Remember, they are one of the organizations that tracks the global temperature.[174]

From: Fred.Spencer@email.com
To: Dialogue@email.com
Subj: No warming?
Date: Friday, September 6, 2013 10:31:15 PM

Are you serious? Why didn't you bring this up a long time ago, Tom? This means we have been discussing all of this for no reason at all. That seems like a pretty bad joke to me.

From: Aileen.Gulledge@email.com
To: Dialogue@email.com
Subj: Not a problem
Date: Sunday, September 8, 2013 4:47:51 PM

OK. I've looked this over and there is no problem here. I will admit I was a little confused at first, but then I sat down and thought it through. We know more energy is coming into the planet than is leaving. Therefore, the temperature has to be going up, even if these charts say otherwise. So, where is the heat going? Then it occurred to me, those charts only show the air temperature and not the temperature of everything else. We already stated that scientist believe the missing heat is going into the oceans. When I did some digging I found some really stunning information about all of this. Just take a look at this chart[175]:

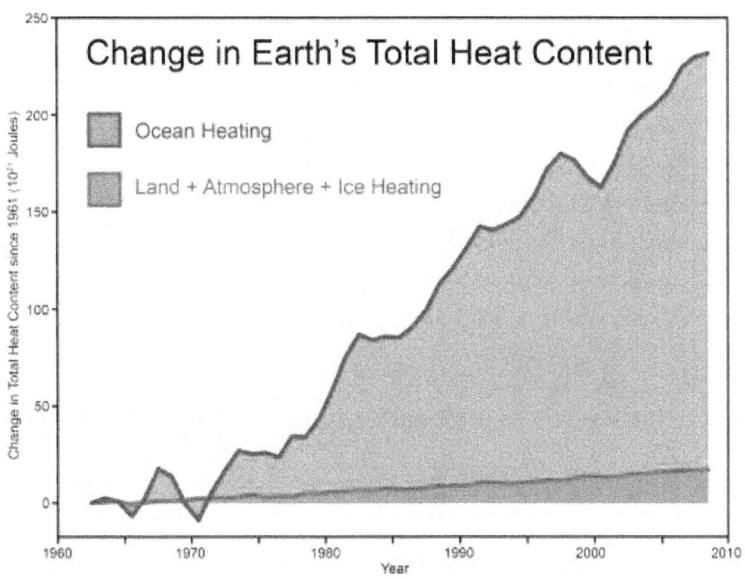

I think this says just about everything that needs to be said on this topic. Yes, the temperature of the atmosphere has plateaued somewhat, but global warming certainly has not slowed down. The oceans are being heated at an alarming rate. So, while the amount of heat stored in the atmosphere hasn't gone up as rapidly as it did before, the amount of heat stored in the oceans has skyrocketed. We can also see that the majority of the global warming actually takes place in the oceans, not the atmosphere. We can safely say that any report that global warming has slowed down is totally false.

From: Fred.Spencer@email.com
To: Dialogue@email.com
Subj: Faked us out
Date: Sunday, September 8, 2013 4:19:51 AM

You knew that, didn't you, Tom?

From: Tom.Lawson@university.edu
To: Dialogue@email.com
Subj: Doing my part
Date: Sunday, September 8, 2013 3:48:11 PM

Of course I did. You don't really think I was going to get blind sided by such a lame claim as that, did you? Aileen hit the real point: when we say global warming we mean the entire globe, not just the atmosphere.

From: Fred.Spencer@email.com
To: Dialogue@email.com
Subj: A little miffed
Date: Monday, September 9, 2013 5:34:13 PM

I still think that was a bad joke. I can just picture you sitting there laughing the whole time.

From: Aileen.Gulledge@email.com
To: Dialogue@email.com
Subj: Just another false argument
Date: Monday, September 9, 2013 1:34:52 PM

I think that might have been the most disturbing one you have pulled out yet, Tom. But, in the end, it was just another false argument. Do these guys have any real science?

From: Tom.Lawson@university.edu
To: Dialogue@email.com

Subj: Denial of science
Date: Tuesday, September 10, 2013 7:12:08 PM

No, I've said it all along. They do not have any real science. They like to claim they do and will talk about 'thousands' of refereed and published papers. They write papers then have each act as referees and then publish them in journals funded by the denier proxies. The papers do not qualify under the scientific method standard and uniformly discredited, but the deniers can then claim they have these published papers to support their claims and some people accept that. All they really have is false arguments and misdirection, lies, half-truths and selected truths to support their claims. This is why I say there is nothing I can say or do to change their minds. They have rejected science and logic in order to get the conclusion they want and all I can do is to produce the science and logic, so I'm powerless to fight their ignorance. And, ignorance is a choice. They have obviously made theirs.

From: Fred.Spencer@email.com
To: Dialogue@email.com
Subj: Urban heat island
Date: Wednesday, September 11, 2013 12:38:47 PM

Now that we are back on track, I have another one that I think could be troubling, although it is another one we have already discussed. The temperature record you keep referring to comes from various weather stations. But, the area around the stations has been built up in recent years. Stations that use to be out in the country are now surrounded by buildings and concrete. This has to make a big difference in the readings and you have to wonder if the change is enough to explain the raising temperatures. I know you addressed this once before, but could you talk about it again?

From: Aileen.Gulledge@email.com
To: Dialogue@email.com
Subj: UHI is not a factor
Date: Friday, September 13, 2013 9:23:36 AM

Sure. I think we really shot that one down, but this is one of those claims that won't go away, no how much science you produce to show it is a false claim. There is just enough credibility to this for it to have some legs.

It is true that the land around some of the stations has changed. Many stations that use to be located in remote locations are now located in urban areas instead. Cities are hotter than country side, so the data has to be checked for bias as a result of extra heat stored in cities. This is called the urban heat island effect.

In fact, a paper by Ross McKitrick, an economics professor at the University of Guelph, and Patrick Michaels, an environmental studies professor at the University of Virginia, concludes that half of the global warming trend from 1980 to 2002 is caused by urban heat island.[176]

Too bad for McKitrick and Michaels that there claim has been shot full of holes. Apparently, they should have done more homework. The first problem with their claim is that the majority of heating observed around the world is actually observed in those remote areas they claim are not warming up. The Arctic tundra, remote areas of Alaska, Canada and Russia, areas far from cities and human development are seeing the greatest amount of heating. Recorded air temperatures in the Arctic region have gone up for every season over the 2000-2009 timeframe with fall and winter seeing the most warming.[177] The southern most area of Canadian tundra has moved northward by about 100 miles in recent decades.[178] In other words, the areas farthest from the city build-up are the areas seeing the biggest increase in temperatures. This is exactly opposite of what would be seen if the heat island effect was the cause of the temperature record. McKitrick and Michaels got it all wrong.

But, there have also been a couple of studies that show the heat island effect is not a problem even for the cases where it is real. One study consisted of removing the urban weather stations and making a new data base. When they compared the two there was hardly any change. And, the small amount of change they saw showed a greater amount of warming when they took the urban stations out. Another study did the opposite. They made a database of weather stations in China from locations that have seen a great deal of buildup over the last 30 years and compared that data base to the original. Again, they did not see any significant deviation. If the heat island effect was a problem then both of these studies should have shown it. Instead, they showed there was basically no change in the data.[179]

But, just in case, another group of scientists made a temperature database that did not use any measurements from any of the stations, or even any air temperature measurements. This, they said, would make a temperature record free of any of the objections the deniers raise. What they found was that their record was pretty much the same as air temperature record recorded from 1901 to 2010. This is one more example that the record of air station measurements is valid.[180,181]

From: Fred.Spencer@email.com
To: Dialogue@email.com
Subj: Scientist deniers
Date: Sunday, September 15, 2013 7:44:10 PM

I read up on some of those papers you cited and I think you made a real good case. It really looks as though the objections to the temperature record are not based in reality.

But, I have been seeing some things that disturb me. There are a couple of scientists that say global warming isn't true and they have published a number of papers to support their claims. One of them is a NASA scientist by the name of Roy Spencer. The other is a professional meteorologist named Anthony Watts.

Also, there is a MIT meteorologist, Richard Lindzen, who is also critical of global warming claims. These are people with credentials and they say it isn't real. If these kinds of people, people who are experts in the field, don't believe in it, then why should any of us lay people?

From: Aileen.Gulledge@email.com
To: Dialogue@email.com
Subj: Richard Lindzen
Date: Monday, September 16, 2013 8:30:24 PM

Wow, Fred! I thought it was Tom's job to do that. But, really, I'm glad you brought these guys up. I have run across their works a number of times. The more you look at them the more you realize they really represent everything wrong about climate change deniers.

Let's start with Richard Lindzen. He is a chaired professor of meteorology at MIT and, as you said, that is an impressive credential. His specialty is to criticize climate change models and he claims they grossly overestimate the effects of increasing levels of greenhouse gases. He concedes that greenhouse gases cause warming, but one of his claims is that Earth will be able to regulate its temperature, like a thermostat, thanks to clouds. His argument is that as the air gets warmer, the amount of moisture in the air will also increase. The moist air will rain out more of its moisture, leaving less to form the wispy, high clouds known as cirrus. Just like greenhouse gases, those cirrus clouds trap heat in the atmosphere, so a decrease in them would counteract the increase of greenhouse gases. As a result, global warming will be controlled.

Basically, his career consists of looking for missing feedback mechanisms. Feedback mechanisms are things that would either make the warming trend get bigger (a positive feedback) or get smaller (a negative feedback). Interestingly, Lindzen spends his time only looking for negative feedback mechanisms. A one-sided, biased approach shows that he is not involved in research as much as he is pursuing a program. He has a record of making his claims in the public forum, only to see his claims discredited. Several times he has come out with some negative feedback mechanism just to see it get disproved.[182] All of his claims disputing climate change have been shot down. His reputation in the professional circles has really been destroyed.[183]

The thing that bothers me the most is the way he lies about how he takes money from the fossil fuel industry. He has stated repeatedly that he takes no money from the fossil fuel industry for his research when, in fact, he takes large amounts of money from the industry. He admitted in one interview that he takes roughly $10,000 per year from the fossil fuel industry in consulting fees. But, the same interviewer found Lindzen charges $2500 per day for his consulting fees and the fossil fuel industry pays for his trips to testify and make speeches, including funds from ExxonMobil. He has also done numerous appearances for The Heartland Institute, and it would be reasonable to believe he takes funding from that organization, as well. My research makes me believe he has taken much more than $10,000 that he claims he took from the fossil fuel industry over the course of his career. Even if he doesn't directly use the funds for his research, he does actually take money from them. The lie of omission is still a lie.

From: Tom.Lawson@university.edu
To: Dialogue@email.com
Subj: Lindzen is a fossil fuel industry tool
Date: Monday, September 16, 2013 9:05:26 PM

OK, you've shown Lindzen is a just a tool for the fossil fuel industry and really doesn't have any credibility. I saw him in a lecture once where he was trying to debunk global warming. His entire lecture consisted of showing errors in the movie, *An Inconvenient Truth*. Many of his arguments were just plain invalid. Yes, there were some errors in the movie, but so what? Even if the entire movie was wrong it doesn't say anything about the science, one way or another. The thing I found most disturbing about his lecture was that when I looked around at the audience, I saw many people nodding their heads in agreement with him. The arguments he made were false, but they seemed credible to the lay person.[184]

Still, what about these other people Fred mentioned?

205

From: Aileen.Gulledge@email.com
To: Dialogue@email.com
Subj: Roy Spencer
Date: Tuesday, September 17, 2013 10:32:38 AM

Let's discuss Roy Spencer next.

Wikipedia has this for an introduction on Spencer, "Roy Warren Spencer is a climatologist, Principal Research Scientist at the University of Alabama in Huntsville, and the U.S. Science Team leader for the Advanced Microwave Scanning Radiometer (AMSR-E) on NASA's Aqua satellite. He has served as Senior Scientist for Climate Studies at NASA's Marshall Space Flight Center. He is known for his satellite-based temperature monitoring work, for which he was awarded the American Meteorological Society's Special Award."[185]

His credentials are real. What he is also known for a series of papers he published in the early 1990s with another scientist, John Christy. In these papers they claimed their analysis of satellite temperature measurements showed no warming in the atmosphere. They became the heroes of the global warming denier crowd as a result.[186]

But, by the end of the 1990s it was shown they were not only wrong, they were so wrong they even admitted it themselves. Separate studies showed four significant flaws in their work. What I find really interesting is that these two are undeniable

206

experts in this process, yet they made four major errors in their data and each of those errors worked to remove global warming.[187] I really have to wonder, what are the chances that two experts will make so many errors and all of those errors work to show the results they wanted? If they were random errors you would expect at least one of them to work to enhance the global warming evidence.

From: Fred.Spencer@email.com
To: Dialogue@email.com
Subj: Faked data?
Date: Tuesday, September 17, 2013 4:18:05 PM

Are you saying they did it on purpose?

From: Aileen.Gulledge@email.com
To: Dialogue@email.com
Subj: John Christy, too
Date: Wednesday, September 18, 2013 10:35:38 AM

No, I can't prove that, but I do have to wonder. And, I think many other people do, too. As they should.

Even though his research has been refuted, Spencer continues to be a global warming denier. There is a whole list of statements by him on the Skeptical Science website with links showing how he is wrong with every one of them.[188] Yes, he is a NASA scientist. But, he is also wrong and that fact is well documented. This is another guy that has damaged his professional reputation.

And, you can throw in John Christy as a freebie. Christy is one of the scientists most frequently quoted by climate change deniers, along with Spencer and Lindzen. Christy actually goes so far as to say the environment is resistant to any manmade greenhouse gas emissions and global warming is merely a naturally occurring cycle. Of course, his 'evidence' for this comes from works by Spencer and Lindzen. It is interesting he keeps making his statements, even though all of his work has been refuted.[189]

From: Fred.Spencer@email.com
To: Dialogue@email.com
Subj: Watts
Date: Wednesday, September 18, 2013 7:57:58 PM

What about Anthony Watts? I know most of the deniers are not scientists and don't really have very much credibility. But, Watts is a professional meteorologist. Like Lindzen, Spencer and Christy, he has professional credentials.

From: Aileen.Gulledge@email.com
To: Dialogue@email.com
Subj: Anthony Watts
Date: Friday, September 20, 2013 2:28:51 PM

I hadn't forgotten him. If we are following the scientific method we need to check out any objections that might be credible.

Watts is the editor for the 'Watts up with that?' (WUWT) blog which is a clearing house for climate change denier articles and has become the number one climate change denial website. The truth is Watts doesn't have the credentials he would like to boast. He uses the title 'meteorologist', but he has no formal education. He went to Purdue, but did not graduate and will not discuss his education. That is a real red flag for me. Watts admits "I'm not a degreed climate scientist" on his blog profile, so at least he isn't being deceitful about it. His primary credential is an American Meteorological Society Seal of Approval, which is not the same as saying he is "AMS Certified" as some sources have inaccurately claimed. The AMS Seal of Approval is a discontinued credential that does not require a bachelor's or higher degree in atmospheric science or meteorology.[190] He was a weatherman for a Fox News AM affiliate station in Chico, California before starting WUWT. Besides his radio duties and WUWT, he founded, and is president of, IntelliWeather Inc. He is also the founder of the Surface Stations Project, which works to document problems with the location of weather stations across the United States.

The goal of the Surface Stations Project is to try and show that the data from the weather stations is invalid due to bad locations for weather stations across the country. Along with the Heartland Institute, he published a glossy report with photos of weather stations he claims are located in bad places and claims of how

209

this invalidated the data. He is one of the main people that keep the urban heat island myth alive.

Of course, we have already discussed that very question and saw that there is no validity to his claims, but let's take a second look anyway because it is very interesting. Scientists at National Oceanic and Atmospheric Administration (NOAA) calculated the average U.S. temperature record both with and without the stations Watts objects to. The results showed almost no difference between the two. What really made it interesting is that the data without the stations actually showed a slightly greater increase in temperature than the data with the stations.[191] Of course, that hasn't stopped Watts from continuing his claims. This illustrates Watts method of operation, and of many climate deniers. They simply make worthless claims that support the conclusion they want and just keep repeating them until people begin to believe they are valid. It isn't an issue if they are proved wrong. Just ignore the proof and keep going. Part of their method is to quote each other, giving the appearance that their claims have been accepted by the scientific community and have validity. That is the primary purpose of his blog and it is no surprise at all that he receives funding from The Heartland Institute.

From: Fred.Spencer@email.com
To: Dialogue@email.com
Subj: Scientists

Date: Friday, September 20, 2013 8:24:31 PM

Any one else?

From: Aileen.Gulledge@email.com
To: Dialogue@email.com
Subj: David Demming
Date: Saturday, September 21, 2013 11:52:28 AM

Yes. Take a look at a news article[192] I saw quoting a number of scientist as being critical of an earlier news report[193] on global warming. They even went so far as to calling it 'hysteria.' The critical article quoted David Deming, a geology professor at the University of Oklahoma. Among the things Demming said was: "The mean global temperature, at least as measured by satellite, is now the same as it was in the year 1980. In the last couple of years sea level has stopped rising. Hurricane and cyclone activity in the northern hemisphere is at a 24-year low and sea ice globally is also the same as it was in 1980."

This is a pretty definite statement and sounds very convincing. But, let's look at the facts for ourselves. He says that the 'mean global temperature, at least as measured by satellite, is now the same as it was in the year 1980." Well, we can check this, and we have. Even the most casual inspection of the data shows Demming's statement is false. He also said, 'In the last couple of years sea level has stopped rising.' The data shows this is simply not true. Take a look at this plot of global sea levels.[194]

211

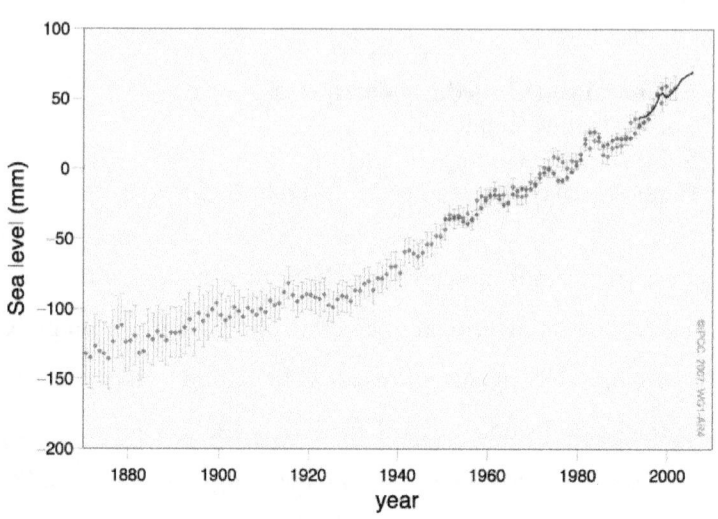

Figure TS.18. Annual averages of the global mean sea level based on reconstructed sea level fields since 1870 (red), tide gauge measurements since 1950 (blue) and satellite altimetry since 1992 (black). Units are in mm relative to the average for 1961 to 1990. Error bars are 90% confidence intervals. {Figure 5.13}

Then, Demming said, 'sea ice globally is also the same as it was in 1980.' Again, as we have already seen, there is no basis in truth for his statement. What about, 'Hurricane and cyclone activity in the northern hemisphere is at a 24-year low'? Actually, the data shows the number of storms has increased and is linked to sea surface warming.[195] This is the pattern. When we make the effort to actually verify statements by the deniers, we see they are completely false.

From: Tom.Lawson@university.edu
To: Dialogue@email.com
Subj: ClimateGate
Date: Saturday, September 21, 2013 3:17:49 PM

But, we have the emails from ClimateGate that shows the climate scientists have been forging the data. They even admit in the emails that climate change isn't real. If the scientists themselves admit it isn't real then there really is a conspiracy going on.

From: Aileen.Gulledge@email.com
To: Dialogue@email.com
Subj: ClimateGate denier hoax
Date: Monday, September 23, 2013 9:53:36 AM

ClimateGate refers to a illegal hacking attack on the Climatic Research Unit (CRU) at the University of East Anglia. On November 17, 2009 someone using an anonymous server in Turkey was able to hack into computers and uploaded a large file. There was also a post by the hacker that read: "We feel that climate science is, in the current situation, too important to be kept under wraps. We hereby release a random selection of correspondence, code, and documents. Hopefully it will give some insight into the science and the people behind it. This is a limited time offer, download now."[196]

However, the emails and documents were not random. The hackers had selectively filtered out emails targeting specific scientists and leaked isolated quotes taken out of context to give the appearance of impropriety on the part of the scientists. The entire incident has been extensively investigated by eight independent committees and each found no evidence of fraud or misconduct.

Digital forensics indicated the hacker was located in eastern North America. The first link to the stolen emails was by climate change denier Stephen McIntyre at climateaudit.com. It has been suggested that McIntyre was involved, but he denies this. What is known, though, is that Saudi web servers were among the first to post the stolen emails and Saudi Arabia was the first country to call for an investigation of the scientists. The lead Saudi climate change negotiator, Mohammed Al-Sabban, quickly used the scandal to help block climate change negotiations at the Copenhagen climate change summit of December 2009 that began just as the stolen emails were released.[197]

So, to be clear, deniers illegally hacked into the CRU computers, stole select emails, quoted them out of context and did it just in time for the climate change summit. Very thorough investigations have completely cleared the scientists of any improper behavior, but deniers keep claiming the opposite. And, it appears there was some involvement by Saudi Arabia, the very same people who were dancing in the streets as the Twin Towers were falling down. Does anyone really think Saudi Arabia has our best interests at heart?

This has actually become an example of how the deniers reject the facts. Despite being completely cleared, the deniers continue to refer to ClimateGate as evidence of misconduct by the scientists. This shows that the deniers are not interested in the truth, they just want to push their agenda and they will use any means available to do so, even if is means lies, deception and character assassination.

From: Fred.Spencer@email.com
To: Dialogue@email.com
Subj: Zingo!
Date: Monday, September 23, 2013 12:24:05 PM

Wow! You really beat them up!

From: Tom.Lawson@university.edu
To: Dialogue@email.com
Subj: Beat em up.
Date: Monday, September 23, 2013 3:43:18 PM

I agree. And, they deserve it. These guys are conducting a well organized disinformation campaign and the result of that campaign is that the environment is being damaged and people are suffering while special interests continue to make piles of money. We are spending our time and resources arguing the issue of whether climate change is occurring instead of working on ways to take care of it. The longer we wait, the harder it will be to fix things. And, these people you mentioned are some of the ones most responsible.

From: Aileen.Gulledge@email.com
To: Dialogue@email.com
Subj: The Heartland Institute
Date: Thursday, September 26, 2013 10:00:13 AM
On the issue of people that are responsible, there is one group I want to add to the list – The Heartland Institute, whom I consider to be a disgusting group of people that are engaged in scientific disinformation for the purpose of monetary gain. There are other professional denier groups, but this group really stood out, in my opinion.

The Heartland Institute is funded by tobacco and fossil fuel interests, which should tell you everything you need to know about them. They, themselves, fund biased conferences on climate change denial and fund climate change deniers. Many of the people that receive their funding have no scientific credentials and many have financial ties to the fossil fuel industry.[198] It appears they are behind many of the disinformation campaigns and attacks on climate scientists. One of their campaigns is to develop school lesson plans to make students believe climate change science is wrong.

Elliot Negin at Huffington Post called the Heartland Institute a 'science free zone' and said, "The Heartland Institute has a long history of peddling half-baked nonsense. Over the years, the

216

group has dismissed the health threat posed by second-hand smoke, ridiculed the evidence for acid rain as "flimflam," and criticized the "hasty phase-out" of chlorofluoro-carbons, which were destroying the atmosphere's ozone layer. More recently, the Economist called Heartland "the world's most prominent think tank promoting skepticism about man-made climate change," and the magazine didn't mean it as a compliment."

Negin goes on to say,

"The documents say that the Charles G. Koch Charitable Foundation contributed $25,000 last year and was expected to contribute $200,000 this year. Mr. Koch is one of two brothers who have been prominent supporters of libertarian causes as well as other charitable endeavors. They control Koch Industries, one of the country's largest private companies and a major oil refiner."

In fact, during the last decade, major publicly traded oil companies have been funding Heartland. If Gillis and Kaufman had dug a little deeper, they would have discovered that ExxonMobil gave Heartland more than $530,000 (2001-2011) while the American Petroleum Institute--the oil and gas industry's trade association--slipped the think tank $25,000 (2008). Two of the Big Three Detroit automakers also chipped in. Chrysler donated $105,000 (2004-2006); General Motors, $165,000 (2004-2010). All had a vested interest in quashing government action on climate change.

An editorial in the Los Angeles Times stated:

> "Leaked documents from the Heartland Institute in Chicago, one of many nonprofits that spread disinformation about climate science in hopes of stalling government action to combat global warming, reveal that it is working on a curriculum for public schools that casts doubt on the work of climatologists worldwide. Heartland officials say one document was a fake, but the curriculum was reportedly discussed in others. According to the New York Times, the curriculum would claim, among other things, that "whether humans are changing the climate is a major scientific controversy."[199]

That is a lie so big that, to quote from *Mein Kampf*, it would be hard for most people to believe that anyone "could have the impudence to distort the truth so infamously." On one side of the "controversy" are credentialed climatologists who publish in reputable, peer-reviewed scientific journals and agree that the planet is warming and that humans are to blame; on the other are fossil-fuel-industry-funded "experts" who tend to have little background in climatology and who publish non-peer-reviewed papers in junk magazines disputing established truths. These are quickly debunked, but not before their findings have been reported by conservative blogs and news outlets, which somehow never get around to mentioning it when these studies are proved to be badly flawed.[200]

In 2012, climate scientist Peter Gleick provided false identity information and got documents on Heartland's donors and its activities. Mark Boslough at Huffinton Post, comparing the Heartland Institute to Al Capone, had this to say, "Like the Market Street Gang, the Heartland Institute has been involved in its own war to control the media. They don't use Tommy guns, but politicians, journalists, and even some scientists are afraid of them. Last year the president of Heartland, Joseph Bast, derided me

219

for exposing his organization's anti-science activities, fabrication of climate data, and defamation campaigns against scientists. Unlike the old Chicago gangs, they don't kill people. But they do engage in character assassination.

Heartland's thugs specialize in smearing scientists on their well-funded website and in the comments sections of newspapers. James M. Taylor is the editor of a Heartland publication, and it is his job to make up fake stuff about scientists. Last year, he was caught falsifying biographical information -- claiming, for example, that Ph.D. scientists don't actually have the degree they list on their resume.

Taylor's long hit list includes high-profile researchers like Prof. Michael Mann of Penn State, whose hockey-stick-shaped temperature graph convincingly showed that global warming is historically unprecedented. Heartland didn't like his science, so they attempted to destroy his career. And now, of course, Peter Gleick is Heartland's public enemy number one. Heartland cannot legitimately address the science, facts, or issues. So they attack individuals.

In the old-style Chicago racketeering days, those who kept their mouths shut were spared from such assaults. But those who spoke out were fair game. The old racketeers burned down stores of shopkeepers who tried to stand up to them. The modern anti-science mob torches reputations, which is the most valuable asset a scientist can own.

After Bast's screed was published in the Santa Fe New Mexican, the Heartland Institute bragged on its website, "The paper used the headline we put on the piece and ran it - word for word - as Bast wrote it. As a former op-ed editor for two newspapers, I can attest that such things are rarely done. Opinion-page staff don't like being told what to do by outsiders..." Bast is a powerful man.

Feb. 14, 2012 may have marked the beginning of the end for Bast and the Heartland Institute, just as it did for Capone in 1929. That was the morning Gleick revealed Heartland's tax forms, lists of donors, and the Board of Directors (including Harrison Schmitt, who withdrew from his controversial appointment to New Mexico's energy and natural resources post in 2011).

The group's political activities outlined in the leaked "fakegate" documents strongly suggest that Heartland's donors may owe back taxes, just like Capone did. Heartland also pays generous stipends to a few scientists and weathercasters who are willing to deny the reality of global warming and generate fake research. Fred Singer, the president of SEPP (another fake science political advocacy group) confirmed to me last year that he gets $5000/month from the Institute, just as the leaked documents say.

Singer, who was in Albuquerque to speak to the fossil-fuel advocacy group, Energy Makes America Great, Inc. ($10 at the door), also spoke to a small science group where the audience was a bit more skeptical. University of New Mexico Prof. David Gutzler, a climate scientist, observed afterward, "...those temperature curves shown by Singer - mostly unlabelled, and all unanalyzed by him - are rising; that's why global warming is considered to be 'unequivocal' by the community that actually does analyze the data."

The documents also revealed that one of Heartland's secret projects is to develop a fake science curriculum for school children, using methods that are effective at dissuading teachers from teaching any science at all. Science teachers who teach factual climate science are next on Heartland's hit list.

Not long after the leak, in an apparent attempt to divert attention, Heartland doubled down by erecting billboards attempting to compare climate scientists to the Unabomber and other murderous criminals. It backfired.

Scientists have long known that the Heartland Institute is a fake science organization. Now we have the documentation to prove it. It's not a think tank. It's a fake tank. Thanks for the Valentine, Heartland!"[201]

I love those editorials and enjoy they way they didn't pull any punches. Obviously, the Heartland Institute is disgusting, but they are only one of many. There are other organizations like them out there. No wonder it has been such an uphill fight. The deniers are not going to be constrained by the law, science or the facts.

223

From: Tom.Lawson@university.edu
To: Dialogue@email.com
Subj: Denier fights
Date: Thursday, September 26, 2013 8:16:48 PM

Yes, the deniers are well funded and well organized. I'm afraid they are winning the battle. The only thing we can hope to do is just keep fighting in the hope that we will eventually win this fight and expose those people for who and what they are.

From: Aileen.Gulledge@email.com
To: Dialogue@email.com
Subj: Skeptical Science
Date: Friday, September 27, 2013 12:20:46 PM

It's a little depressing. But, one thing I have learned doing all of this is that the science is there for anyone to see and it is overwhelming. Hopefully, more and more people will do their own homework and see the truth of what is going on.

It would be difficult for us to cover every wild claim made by the deniers or to examine each of the main characters in the denial campaign. But, it is possible with only a little effort to research any claim made. One of the best debunking sites is the Skeptical Science website and their Global Warming and Climate Change Myths webpage at http://www.skepticalscience.com/argument.php. I recommend it as a resource for dealing with all of the denial claims.

From: Fred.Spencer@email.com
To: Dialogue@email.com
Subj: Shooting the messenger?
Date: Friday, September 27, 2013 5:38:07 PM

Aren't you just shooting the messenger, Aileen?

From: Aileen.Gulledge@email.com
To: Dialogue@email.com
Subj: Just showing the methods used
Date: Saturday, September 28, 2013 8:28:20 AM

No, that isn't what I'm trying to do. I have shown that each of these scientists has a track record of making false claims and that is a valid point to make. When someone goes into the public forum and makes claims, what they have said or done in the past is a pattern of behavior that is something to consider. If someone has a long track record of being meticulous in their research and careful about what they say, it is different than if they have a history of making crazy, unsubstantiated claims and wild accusations. The character and motivations of the individuals is something that is a valid part of the message.

From: Fred.Spencer@email.com
To: Dialogue@email.com
Subj: Scientists track records
Date: Saturday, September 28, 2013 7:05:22 PM

But, doesn't that go both ways? If we are going to expose the track history of the deniers, can't they do the same to the climate change advocates?

From: Aileen.Gulledge@email.com
To: Dialogue@email.com
Subj: Must be responsible
Date: Sunday, September 29, 2013 2:45:33 PM

They not only can do that, but they should. Of course, that will typically go against them because the history of these scientists is usually excellent. Keep in mind, there is a difference in looking at the track record and engaging in character assassination. That is, unfortunately, something the deniers do and that does nothing constructive. It is only hate speech.

From: Tom.Lawson@university.edu
To: Dialogue@email.com
Subj: Debate the issues
Date: Monday, September 30, 2013 3:41:40 PM

The deniers say they only want a debate on the subject and that they are being locked out. The claim the climate change advocates aren't interested in even considering any alternative explanations.

From: Aileen.Gulledge@email.com
To: Dialogue@email.com
Subj: Plenty of debate
Date: Tuesday, October 1, 2013 8:18:07 PM

Any claim there is a lack of debate is clearly a false one. Just look at what we have been discussing here. We have looked at several scientific claims made by scientists that deny climate change. For instance, we have discussed claims by Richard Lindzen, John Christy, Roy Spencer and David Demming. Their claims were all presented and examined via the scientific method and all of their claims were found to be invalid. Additionally, the claims of Anthony Watts concerning the heat island effect have also been examined by the scientific community. Keep in mind that Watts does not have any scientific credentials, and yet, the scientific community treated his claims seriously and investigated them, just to find they were completely wrong.

There are more deniers making claims than what we have discussed here, and their claims get examined via the scientific method. So, here we see the very debate they demand going on, just as it should. But, having a debate doesn't mean their claims have to be accepted. They get a chance to have their say and the scientific community investigates their claims. The reason they say there isn't a debate is because their claims all get rejected. The reason their claims get rejected is because they don't have any science to back up their claims.

Having a debate doesn't mean their claims get accepted, it means they get examined. Of course, we have seen that is always the fatal flaw in the claims by the deniers. They can never stand up to examination.

Without exception, their claims have been proven to be invalid.

And yet, all of these people keep making their claims and the deniers keep referring to them. How is it that people just ignore the science and keep going as if nothing has happened?

From: Tom.Lawson@university.edu
To: Dialogue@email.com
Subj: Rejecting science
Date: Tuesday, October 1, 2013 10:08:05 PM

Remember what I said, the only way you can ignore the reality of global warming is by rejecting science. Each of these people, and the people that follow them, have all rejected science in favor of finding evidence that supports what they want to believe, even if that evidence is invalid.

From: Aileen.Gulledge@email.com
To: Dialogue@email.com
Subj: I'm a believer!
Date: Wednesday, October 2, 2013 9:01:40 AM

I thought you were exaggerating when you said that before, Tom. But, I believe you now. The science is so overwhelming the only way you can deny global warming is to deny science.

From: Fred.Spencer@email.com
To: Dialogue@email.com
Subj: Proven?
Date: Wednesday, October 2, 2013 4:16:27 PM
So, Tom, did we prove manmade global warming is real via the scientific method?

From: Tom.Lawson@university.edu
To: Dialogue@email.com
Subj: Proven!
Date: Wednesday, October 2, 2013 5:59:10 PM

I feel very satisfied.

From: Aileen.Gulledge@email.com
To: Dialogue@email.com
Subj: Overwhelming evidence
Date: Thursday, October 3, 2013 1:27:27 PM

And, Tom, I agree with what you said at the beginning. The
amount of information I found doing this was overwhelming.
The problem wasn't proving global warming, the problem was
limiting myself to what we talked about. We could review
information about this for the rest of our lives and not go through
it all. There is some new report literally every day. You are
right, there is so much out there that anyone can prove it now.
The only way anyone could still deny global warming is because
they want to.

From: Fred.Spencer@email.com
To: Dialogue@email.com
Subj: Selling the message
Date: Thursday, October 3, 2013 9:13:05 PM

And, that is always the problem with marketing. You have to
deliver the message the customer wants. You cannot sell
something, no matter how good it is, if the customers have made

up their minds they don'want it. The problem is finding a way to make the customer want the message of global warming and what to do about it. By the way, you haven't discussed that point.

From: Aileen.Gulledge@email.com
To: Dialogue@email.com
Subj: Been busy
Date: Friday, October 4, 2013 10:12:40 AM

I've been a little busy proving the world needs saving. I haven't had time to save it, also.

From: Tom.Lawson@university.edu
To: Dialogue@email.com
Subj: Let's save the world
Date: Friday, October 4, 2013 5:15:24 PM

Well then, maybe its time.

"Illusions commend themselves to us because they save us pain and allow us to enjoy pleasure instead. We must therefore accept it without complaint when they sometimes collide with a bit of reality against which they are dashed to pieces."

– Sigmund Freud

Chapter 7 – What Can Be Done?

From: Tom.Lawson@university.edu
To: Dialogue@email.com
Subj: For what purpose?
Date: Saturday, October 5, 2013 9:24:15 AM

I think that was a good job. You gave a nice, air-tight argument and addressed the claims made by the deniers. However, I don't think it is any real use.

From: Fred.Spencer@email.com
To: Dialogue@email.com
Subj: Re: For what purpose?
Date: Saturday, October 5, 2013 3:31:58 PM

What do you mean?

From: Tom.Lawson@university.edu
To: Dialogue@email.com
Subj: Denier logic
Date: Saturday, October 5, 2013 7:47:33 PM

People that believe in manmade global warming don't need us to prove it to them. They have looked at the evidence for themselves and reached a logical conclusion. People that don't believe never will. They have rejected science and logic in order to reach the conclusion they want to come to. There is no amount of scientific proof or logical argument that will ever convince them to change their minds. They could be standing in boiling water with their hair on fire and they will still insist it is just a normal, natural cycle.

From: Aileen.Gulledge@email.com
To: Dialogue@email.com
Subj: Or lack thereof
Date: Sunday, October 6, 2013 2:08:36 PM

Maybe. But, we can still try.

From: Fred.Spencer@email.com
To: Dialogue@email.com
Subj: Some are undecided
Date: Sunday, October 6, 2013 6:43:46 PM

And, Tom, don't forget about the people that haven't decided yet. There is a large percentage of the population that is still wondering which side of the argument to believe in. If we just sit back and let the other side spread their nonsense without ever objecting they will win by default. By publicizing the science and showing the inability of the deniers to prove any of their claims we can persuade some of the undecided to believe in the science. Certainly, the more we can show how baseless the claims made by deniers are, the more likely it is that rationale people will reject their arguments altogether.

From: Tom.Lawson@university.edu
To: Dialogue@email.com
Subj: Some will learn
Date: Sunday, October 6, 2013 7:57:38 PM
Of course, you are both right. That is why I'm in the business. Not everyone will learn and believe, but some will. If we can convince enough then we can stop spending our time arguing about if it is real or not and start doing something about it. Which brings me to one last question, now that you proved global warming is real, what do you propose we do about it?

From: Aileen.Gulledge@email.com
To: Dialogue@email.com
Subj: Fix global warming?
Date: Sunday, October 6, 2013 9:19:15 PM

You mean, what do I propose to do about fixing global warming?

From: Tom.Lawson@university.edu
To: Dialogue@email.com
Subj: Can't be fixed
Date: Sunday, October 6, 2013 10:12:33 PM

Yep! Now that you have shown that it is real, what do you think we should do about it? Why should we get upset about something that can't be fixed? The technology isn't there and,

233

even if we could do something about it, we would break the bank paying for it and it would result in lowering our standard of living. Maybe we should just get use to the idea of global warming and deal with it.

From: Fred.Spencer@email.com
To: Dialogue@email.com
Subj: Don't want to return to the stone ages
Date: Monday, October 7, 2013 4:00:41 AM

This is a big one, Aileen. I am not interested in going back to the stone ages, even if man made climate change is real. If people are saying to give up our cars and walk, I won't listen. I am not going to give up my air conditioning, either. Or, my electronics. And, I'm not turning the lights out, either. If the only way we can save the world is by destroying civilization, then I say we take our chances.

And, most of the world feels the same way. The key to a standard of living is electricity. Most of the world has a much lower standard of living than we do and would like to have a better one. They are willing to go through a lot in order to get there. How much are people willing to put up with? Just do an Internet search for images on 'China, air pollution' and take a look at just what people are willing to live with if you give them a job and a higher standard of living.

Between them, China and India have nearly one-half of the

world's population and they are committed to improving their standards of living. In order to do that they need to build more power plants. That doesn't even include what is going on in Africa and South America. Right now, there are about 1200 coal-fired power plants in various stages of production across 59 different countries.[202] Why should we in America lower our standard of living when the increase in emissions from the rest of the world is many times what we could possibly cut?

From: Aileen.Gulledge@email.com
To: Dialogue@email.com
Subj: I don't want the stone ages either
Date: Monday, October 7, 2013 12:32:12 PM

Wow! I'll have to get back to you.

From: Aileen.Gulledge@email.com
To: Dialogue@email.com
Subj: Options are there
Date: Monday, October 14, 2013 9:53:24 AM

All right. It wasn't as bad as I expected. There is actually quite a bit of literature on the subject and a lot of serious research has been done. The IPCC claims there are technologies in operation or in the pilot stage that could keep carbon dioxide levels from doubling over the next hundred years.[203] But, this was refuted in a paper that appeared in the journal *Science*[204] where the authors said the IPCC overstated the issue and we need to invest in a revolutionary change in technology. However, from what I have read it is pretty obvious we could do a lot about global warming, if we really wanted to. And, I mean we could do a lot without paying gigantic bills or lowering our standard of living. And, it could be done with technology that is available today. It is really all about just making the decision to do it.

From: Tom.Lawson@university.edu
To: Dialogue@email.com
Subj: Re: Options are there
Date: Monday, October 14, 2013 5:14:04 PM

Bold statement!

From: Aileen.Gulledge@email.com
To: Dialogue@email.com
Subj: 15 strategies
Date: Tuesday, October 15, 2013 9:23:53 AM

Yes, but one that is supported by the evidence. Just like the global warming claims themselves.
Basically, we need to reduce global greenhouse gas emissions by 50% by the year 2050. This sounds drastic at first, but the number show it is easily within reach with today's technology. If you look at the amount of emissions per person by country, the European countries emit about one-half as much as we do in the U.S.[205] and I don't think anyone would argue that Europeans suffer from a lower standard of living. In fact, in many ways, they enjoy a superior standard of living than we do.

There was a paper in *Science* where the authors studied this question and listed 15 strategies that are available right now that would reduce carbon emissions.[206] One of the great things is this paper was published in 2004. The technologies are even better developed today. What they looked at is the question of how to keep the CO_2 levels steady for the next 50 years. They left the question of reducing levels beyond that to research and development of new technologies.

From: Fred.Spencer@email.com
To: Dialogue@email.com
Subj: Re: 15 strategies
Date: Tuesday, October 15, 2013 6:12:48 PM

Tell us about these strategies.

From: Aileen.Gulledge@email.com
To: Dialogue@email.com
Subj: More efficient cars
Date: Thursday, October 17, 2013 5:25:50 PM

The first one is to make automobiles more efficient. They use the figure of doubling the efficiency of cars from 30 miles to the gallon to 60. With the anticipated growth of automobiles to 2 billion cars worldwide, they estimate this would result in about one-seventh of the needed cuts. A companion strategy was to cut vehicle use in half. Again, with the expected growth of automobiles worldwide, they estimate this would result in an equal amount of carbon dioxide reduction. I'm not sure how feasible it is to double the average mileage of cars worldwide and to cut worldwide usage in half at the same time, but any progress in either would have the potential to result in very large reductions.

From: Fred.Spencer@email.com
To: Dialogue@email.com
Subj: Doubling mileage and cutting usage in half
Date: Thursday, October 17, 2013 9:43:23 AM

Doubling the mileage of American cars will be tough, but doable. We like gas-guzzlers, so there is plenty of room for improvement. The White House has, in fact, issued guidelines to increase the mileage of American cars to 54.5 mpg, almost double of what it is today.[207] Studies by the National Highway Traffic Safety Administration and the EPA estimate it will cost about $157 billion to implement, but save between $419 and $515 billion in reduced fuel costs.[208] But, there are going to be some problems they don't talk about. People keep cars for longer nowadays. So, someone buying a car that conforms to older standards will be driving that car long after the new standards take affect. The point being that the entire U.S. fleet of automobiles will not just suddenly be getting twice the mileage it use to. It will take time for it to occur and that reduces the total amount of cuts in carbon dioxide.

While we can do it in this country, I'm not as sure about doing it in other countries where the average mileage is already much higher than here in the U.S. Gasoline is much more expensive in other countries and they drive much more efficient cars. I'm not sure how much we could improve on the mileage in these other countries. But, of course, any improvement is good.

Reducing the amount of vehicle usage in half is something that would be very difficult to achieve in America. We are married to our cars and don't want to give them up. Public transportation is very poor in most places and does not satisfy the needs of most people. There is room for improvement, but cutting usage in half will be tough.

I also think it would be tough to cut vehicle usage in half in other countries. They already have well-established public transportation that is heavily used. Improvements in the systems can increase how much they are used, but I don't see usage doubling. More efficient planning will help some, but people will still need to go to work and back and that is a significant amount of the driving that is done. We can reduce the usage, but I don't see a worldwide 50% reduction in the average distance cars are driven every year.

From: Tom.Lawson@university.edu
To: Dialogue@email.com
Subj: Can still result in emission reductions?
Date: Saturday, October 19, 2013 2:44:46 PM

But, you still agree we can reduce carbon emissions this way?

From: Fred.Spencer@email.com
To: Dialogue@email.com
Subj: Reductions in emissions possible
Date: Saturday, October 19, 2013 5:32:30 PM

Yes! Don't get me wrong. I think both are a great idea and we should pursue them. I think they will result in significant reductions of carbon emissions. I just don't think they will result in as much reduction as claimed.

From: Aileen.Gulledge@email.com
To: Dialogue@email.com
Subj: More efficient buildings
Date: Saturday, October 19, 2013 7:01:18 PM

The next strategy they propose is making buildings more efficient. Making buildings more efficient includes things like better insulation, better water heating, lighting and refrigeration. Again, they estimate about one-seventh of the needed reductions could result from these kinds of changes. Interestingly, they say half of the worldwide reductions would come from the developing countries.

From: Fred.Spencer@email.com
To: Dialogue@email.com
Subj: New buildings are more efficient
Date: Sunday, October 20, 2013 3:40:56 PM

That is probably because most of the construction of new

buildings is likely to occur in the developing countries. The developed countries have existing buildings that are not going to be changed easily. Of course, the lighting will improve as we get off incandescent bulbs and go to more efficient lighting like compact fluorescent lights and LED lights. Improvements that are being made in appliance efficiency is impressive. Things that consume electricity, such as refrigerators, water heaters and clothes dryers are seeing tremendous improvement in efficiency. I think we will see some real energy savings in even the developed countries in this area.

From: Aileen.Gulledge@email.com
To: Dialogue@email.com
Subj: Power plant efficiency
Date: Monday, October 21, 2013 8:48:39 PM

The next one is to improve power plant efficiency. They discuss the saving of coal-fired power plants improving their efficiency from 32% in 2000 to 60% in 2054. This would result in one-seventh of the needed reductions.

From: Fred.Spencer@email.com
To: Dialogue@email.com
Subj: Double the efficiency?
Date: Monday, October 21, 2013 9:20:27 PM

Improving efficiency is something that can be done, and needs to be done. But doubling it? That is a tall order. Efficiency can be increased by operating at full capacity instead of having to change output during the day. Also, there are places where energy is wasted, such as not operating at the most optimum temperature. If those areas are taken care of, efficiency goes up. I don't see efficiency doubling, though. I think a greater source of carbon reduction would be to just not build as many coal plants. I know coal is an important part of the total energy package, but there are alternatives that don't produce as much carbon. And, with increased regulations, coal-fired plants are becoming more expensive.

From: Aileen.Gulledge@email.com
To: Dialogue@email.com
Subj: Decarbonization
Date: Wednesday, October 23, 2013 8:48:39 PM

That is topic of a couple of their strategies: decarbonization of power plants and substituting natural gas for coal.

From: Fred.Spencer@email.com
To: Dialogue@email.com
Subj: Reduce the number of coal fired power plants
Date: Thursday, October 24, 2013 6:46:43 PM

That is an area I would expect to see some significant improvements. China plans to start a new coal fired power plant every week until 2030 to meet growing energy demands. If they can be convinced to use alternative sources the savings in carbon emissions would be huge. Even here in America, our energy demand is skyrocketing. We need to build other sources of

242

energy than just coal. Nuclear power is the single best option. It is unfortunate that the environmental group has polluted the public opinion of nuclear power because it is, by far, the most environmentally friendly source of power. We could greatly reduce our carbon emissions with the construction of new nuclear plants with their modern designs. People talk about the potential for accidents without mentioning the potential of accidents at other plants, especially coal-fired plants. The sludge from those plants is horribly toxic and will remain toxic for all eternity. Gigantic amounts are stored at the power plants, just waiting for accidents to occur.

In fact, according to the EPA, U.S. emissions of greenhouse gases has been dropping for many years and is now to the levels of about 1994. This is very encouraging news. Much of this reduction has been from using natural gas instead of coal in power plants.[209]

So, yes, there is great potential to reduce carbon emissions this way.

From: Aileen.Gulledge@email.com
To: Dialogue@email.com
Subj: Wind and solar power
Date: Friday, October 25, 2013 1:07:15 PM

They also mention wind power and solar power.

From: Fred.Spencer@email.com
To: Dialogue@email.com
Subj: Green energy
Date: Friday, October 25, 2013 4:59:37 PM

These are good, but they are really capable of providing only low amounts of energy. To give you a comparison, a wind farm covering an area the size of Rhode Island can produce, at best, the amount of power generated by one nuclear power plant. I think there is all sorts of opportunities here, especially for off-shore wind farms, but I don't think they will make as much of a dent in the carbon emissions as some advocates like to claim.

Solar cells could make a pretty good dent, but they are very expensive. To equip a house with enough solar cells to make them energy independent is so expensive it takes about eight years of energy savings to pay for it. Granted, after that it is money in the bank, but you have to find a way to pay for it in the first place.

Still, solar cells, especially as they become cheaper and more efficient, have the potential to really cut into carbon emissions.

From: Aileen.Gulledge@email.com
To: Dialogue@email.com
Subj: Hydrogen power
Date: Sunday, October 27, 2013 4:08:54 PM

How about hydrogen power? That is another of the options they discussed.

From: Fred.Spencer@email.com
To: Dialogue@email.com
Subj: H-powered cars
Date: Sunday, October 27, 2013 8:09:33 PM

At one time I had a dim view of the future of hydrogen powered cars. But, I met with some engineers from General Motors about their hydrogen powered SUV. It was very cool. It was a prototype model, but they put 1000 of these in people's homes and let them drive them for three months. Then, they put them in other people's homes. So, they have built up a good data base on the performance and they said that it has performed very well. They hope to have a production model by 2015. It carries 4 kilograms of hydrogen and that is enough to go about 200 miles. The cost of the H is about $1.50 per kilogram. So, you can drive 200 miles for about $6. Obviously, you better have a fueling station within 200 miles, which we currently do not. have But, they told me for about $14 billion dollars, we can put H-stations on the busiest traffic routes and this would put a refueling station within 2 miles of over 90% of the population. Of course, this would require the government to put up the money, but it is really quite doable.[210]

Cars represent 70% of our oil consumption. So, if we got our cars off oil, we make a big reduction in carbon emissions.

My thought on how to make it work is to start with city fleets. They represent a captive customer that isn't driving all over the country and it would be possible to provide fueling stations for them. For instance, school buses all go to the same yard for refueling. As the fleet grows and more H-stations are provided, you start to make them open for public use, this provides the ability for individuals to buy H-powered cars and know they can refuel them. Then, as more and more cars are on the streets, H-stations start to spread. After a while, you have the necessary infrastructure to support these cars all over the country.

One thing that I thought was a real show stopper is the supply of hydrogen. I always thought we didn't have hydrogen available, just hydrogen compounds, and it takes a lot of energy to free up hydrogen from its compounds. But, the GM engineers insist that there is lots of hydrogen coming out of oil and natural gas wells and we are just burning it off. I don't know how much that is, but that would make a big difference in the availability of raw hydrogen.

I really think this is one option that could make a huge difference. And, the technology is there.

From: Aileen.Gulledge@email.com
To: Dialogue@email.com
Subj: Grow more trees
Date: Sunday, October 27, 2013 10:08:38 PM

How about reducing deforestation and planting new forests?

From: Fred.Spencer@email.com
To: Dialogue@email.com
Subj: Forests
Date: Tuesday, October 29, 2013 2:02:34 PM

I think that is something that will help, but it will never solve the
problem. And, I think it would only be a temporary help. Yes,
the growth of trees and other plants will remove carbon from the
atmosphere, but it would take massive amounts of land to
accomplish significant reductions in atmospheric carbon. Then,
once the forests grow to maturity there will be routine die-off
that will return the carbon back to the atmosphere as quickly as it
is taken in. At some point, it will become a neutral factor. But, it
will help in the short run.

But, you didn't say anything about cap and trade or about a
carbon tax, Aileen.

From: Aileen.Gulledge@email.com
To: Dialogue@email.com
Subj: Cap and trade
Date: Friday, November 1, 2013 6:08:16 PM

I was coming to those. Let's talk about the cap and trade
program first.

The idea of cap and trade is pretty simple. A cap is established,
'using science', on how much can be emitted in the way of

247

greenhouse gases. Allowances, totaling up to this limit, are then auctioned off or assigned to entities that emit greenhouse gases. Anyone exceeding their allowance is subject to a penalty. Then, the total amount of emissions is lowered every year. The plan is that this lowering will occur on a known schedule so that companies can plan ahead. Then, if a company doesn't use its quota, they can trade or sell that excess to someone else.[211] This is suppose to make it more economical for them to lower their greenhouse gas emissions and give them an incentive to find better ways to do whatever it is they do. This idea worked very well about 20 years ago in reducing sulfur dioxide emission responsible for acid rain.[212]

From: Fred.Spencer@email.com
To: Dialogue@email.com
Subj: Cap and trade problems
Date: Saturday, November 2, 2013 7:04:42 PM

Of course, things are different when you raise the stakes. There are a number of problems with cap and trade. The first problem is that when Congress was working on a bill to implement this plan they handed out numerous exemptions to win over critics. In fact, they handed out so many that the worry was the bill would just become a 'cap and tax' program that would kill the economy without reducing greenhouse emissions. The bill died a quiet death in Congress and no one is willing to even mention it any more.[213]

Another issue is that it is believed companies with captive audiences, such as power companies, would simply pay the fines and then pass that cost on to the consumers who would have no choice but to pay the higher cost.[214]

From: Aileen.Gulledge@email.com
To: Dialogue@email.com
Subj: Carbon tax
Date: Sunday, November 3, 2013 1:27:23 PM

And, that brings us to the carbon tax.

The idea behind a carbon tax is that people that emit greenhouse gases do not pay their fair share of the expenses that result from their actions. Someone buys gasoline and drives their car around. They pay for the gas, but the emissions from the car contribute to global warming and they have done nothing to pay for the damages resulting from global warming. So, a carbon tax would be added to the cost, in this case, the gasoline so that some of the resulting costs can be covered.[215]

There would be two effects from a carbon tax. The first would be additional revenue for the government. Since it is the government that will have to shoulder the majority of the cost of global warming, the idea is that this additional revenue will merely cover the additional expenses the government has to cover because of global warming.

The second effect would be to make it more economical to use alternative energy sources that don't produce greenhouse gases. This would work to reduce the total amount of greenhouse gas emissions.

From: Fred.Spencer@email.com
To: Dialogue@email.com
Subj: ExxonMobil
Date: Sunday, November 3, 2013 9:35:30 PM

Strangely, one of the supporters of the carbon tax is ExxonMobil. The reason I say this is strange is because they would have been one of the big winners under cap and trade. But, on their website, they state,

> If policymakers do move to impose a cost on carbon, we believe that a carbon tax would be a more effective policy option to reduce greenhouse-gas emissions than alternatives such as cap-and-trade. And to ensure revenues raised from such a tax are indeed directed to investment, and to assist those on lower incomes who spend a higher proportion of their income on energy, a carbon tax should be offset by tax reductions in other areas to become revenue neutral for government. It is rare that a business lends its support to new taxes. But in this case, given the risk-management challenges we face and the

policy alternatives under consideration, it is our judgment that a carbon tax is a preferred course of public policy action versus cap and trade approaches.[216]

It makes it sound as though ExxonMobil really is changing its policy, until you remember they are still funding The Heartland Institute and the American Petroleum Institute. Words and actions still don't match with them.

From: Tom.Lawson@university.edu
To: Dialogue@email.com
Subj: Like a knife through butter
Date: Sunday, November 3, 2013 10:07:46 PM

Wow, Fred, you went right through those.

From: Fred.Spencer@email.com
To: Dialogue@email.com
Subj: Part of my life
Date: Monday, November 4, 2013 5:45:20 AM

I need to keep up on this stuff for my business. I read about these things almost every day.

From: Tom.Lawson@university.edu
To: Dialogue@email.com
Subj: Business affected?
Date: Monday, November 4, 2013 7:31:16 PM

Now, that is an interesting comment. What you're saying is climate change affects business.

From: Fred.Spencer@email.com
To: Dialogue@email.com
Subj: The cost to business
Date: Tuesday, November 5, 2013 3:35:34 AM

Absolutely. Nationwide, we are talking about hundreds of billions of dollars of damage costs due to climate change. Worldwide, the estimate is in the trillions of dollars. And, I don't mean some time in the future, it is happening right now. The DARA Group is a nongovernmental organization based in Europe. They were commissioned by 20 governments to assess the damage due to climate change. They concluded climate change is already contributing to the death of about 400,000 people and costing the world $1.2 trillion a year, wiping out about 1.6% of the world's GDP.[217] That number is only going to rise over time.

From: Aileen.Gulledge@email.com
To: Dialogue@email.com
Subj: Joking?
Date: Tuesday, November 5, 2013 10:30:49 AM

Trillion? Was that a typo?

From: Fred.Spencer@email.com
To: Dialogue@email.com
Subj: Poor hit hardest

Date: Tuesday, November 5, 2013 6:13:21 PM

No. And, under-developed countries will suffer more than the developed ones. We are looking at massive realignments of world economies over the next 100 years. You can only speculate about what will happen to world tensions.

From: Aileen.Gulledge@email.com
To: Dialogue@email.com
Subj: Any hope?
Date: Wednesday, November 6, 2013 10:50:17 AM

But, you agree there is something we can do about it?

From: Fred.Spencer@email.com
To: Dialogue@email.com
Subj: $10,000 Global Warming Skeptic Challenge
Date: Thursday, November 7, 2013 5:44:28 AM

Yes, that list you went through would have a significant impact on greenhouse gas emissions, if it was implemented. And, the things you listed are not everything. It shows we can do something about climate change if we decide to and it will not bankrupt the economy or lower the standards of living. Just like everything else the deniers say, it is not based on the facts.

In fact, I am so satisfied with Aileen's work that I am throwing out a couple of challenges to climate change deniers. If they can prove, through the scientific method, that manmade climate change is not real, I will pay them $10,000. Let's call it the $10,000 Global Warming Skeptic Challenge.

From: Tom.Lawson@university.edu
To: Dialogue@email.com
Subj: Rules?
Date: Thursday, November 7, 2013 6:03:10 PM

Pretty bold, Fred. What are the rules?

From: Fred.Spencer@email.com
To: Dialogue@email.com
Subj: Entry rules
Date: Thursday, November 7, 2013 8:00:39 PM

Pretty simple.

1. I will award $10,000 of my own money to anyone that can prove, via the scientific method, that man-made global climate change is not occurring;

2. There is no entry fee;

3. You must be 18 years old or older to enter;

4. Entries do not have to be original, they only need to be first;

5. I am the final judge of all entries but will provide my comments on why any entry fails to prove the point.

That's it! I know no one is going to get rich with $10,000. But, you would think some denier out there would like to have a spare $10,000. After all, the deniers all claim it is a simple matter, and it doesn't even have to be original. If it is so easy, just cut and paste the proof from somewhere. Provide the scientific evidence and prove your point and the $10,000 is yours!

Anyone that is interested can make a submission at dialoguesonglobalwarming.blogspot.com.

But, I am sure I will never have to pay because it can't be proven. The scientific evidence for global warming is overwhelming and no one can prove otherwise.

From: Aileen.Gulledge@email.com
To: Dialogue@email.com
Subj: Maybe go both ways?
Date: Friday, November 8, 2013 10:35:04 AM

I don't know Fred. I just proved it is real. Maybe I can go back and prove it isn't real.

From: Fred.Spencer@email.com
To: Dialogue@email.com
Subj: Money is waiting to be claimed
Date: Friday, November 8, 2013 4:31:46 PM

If you can, the money is yours.

From: Tom.Lawson@university.edu
To: Dialogue@email.com
Subj: Other challenge?
Date: Friday, November 8, 2013 8:04:45 PM

But, you mentioned there were a couple of challenges. What's the other challenge?

From: Fred.Spencer@email.com
To: Dialogue@email.com
Subj: $1000 Scientific Evidence Challenge
Date: Saturday, November 9, 2013 3:28:15 PM

Our discussion has convinced me that not only are the deniers wrong, but they actually have no scientific evidence at all to support their claims, and I'm willing to state just that. If the deniers have ANY scientific evidence at all, then I will give them a chance to prove it and will pay them if they can. So, I am also setting up the $1000 Scientific Evidence Challenge. The rules are the same, except that any submissions have to provide scientific evidence that manmade climate change is not real. It does not have to prove climate change is not correct, that is the test of the first challenge. This challenge only requires that someone show there is any scientific evidence at all to support the claims of the deniers.

Anyone that wants to take the challenge can go to the same website to make a submission, dialoguesonglobalwarming.blogspot.com.

I really do believe there is absolutely no scientific evidence to support them. Now, I'm putting my money where my mouth is.

From: Tom.Lawson@university.edu
To: Dialogue@email.com
Subj: Safe bet
Date: Saturday, November 9, 2013 6:05:08 AM

I think your money is safe. I don't think even Aileen could do it. But, if someone does, I will change my beliefs. I believe in the science and I will go where it takes me. Right now, I am very comfortable where it is taking us.

From: Fred.Spencer@email.com
To: Dialogue@email.com
Subj: Undeniable
Date: Saturday, November 9, 2013 8:38:25 PM

I agree. The evidence is undeniable.

Index

259

End Notes

Chapter 1 – Why we should care

[1] Pappas, Stephanie, How Scientists Cope as Climate Debate Gets Personal, Live Science, http://www.livescience.com/17257-scientists-cope-personal-climate-debate.html, Dec 1, 2011

[2] Mustain, Andrea, Sticking it to the Mann: Climate science sparks protest, Live Science, http://www.livescience.com/16789-mann-climate-science-sparks-protest.html, Oct 31, 2011

[3] *ibid.* 2

[4] True story

[5] Feder, Toni, Climate scientists not cowed by relentless climate change deniers, *Physics Today*, Vol. 65, No. 2, Feb 2012, pg 22

[6] *ibid. Feder*

[7] *ibid.* 1

[8] *ibid.* 1

[9] Somerville, Richard C. J. and Susan Jay Hassal, Communicating the science of climate change, *Physics Today*, *Vol 64*, No. 10, October 2011, pg 48

[10] Demelle, Brendan, Heartland Institute Exposed: Internal Documents Unmask Heart of Climate Denial Machine, Desmogblog.com, http://www.desmogblog.com/heartland-institute-exposed-internal-documents-unmask-heart-climate-denial-machine, Feb 14, 2012

[11] ZME Science, 9 out of 10 top climate change deniers linked with Exxon Mobil, http://www.zmescience.com/ecology/climate-change-papers-exxon-mobil/, May 10, 2011

[12] Butler, E.M., *The Myth of the Magus*, The Press Syndicate of the University of Cambridge, 1948, pg 11

[13] Paine, Thomas, *The American Crisis*, 1777

[14] True story

[15] Cullen, Heidi, *The Weather of the Future, Heat Waves, Extreme Storms, and Other Scenes from a Climate-Changed Planet*, Harper, NY, pg 5

[16] Maxfield, John, *Hurricane Sandy, one year later: Assessing the economic cost*, Huffington Post, http://www.dailyfinance.com/on/hurricane-sandy-anniversary-economic-cost/, Oct 26, 2013

[17] Romm, Joe, *Superstorm Sandy's link to climate change: 'The case has strengthened' says researcher*, Climate Progress, http://thinkprogress.org/climate/2013/10/28/2843871/superstorm-sandy-climate-change/, Oct 28, 2013

[18] Biello, David, Climate change has helped bring down cultures, Scientific American, http://blogs.scientificamerican.com/observations/2012/01/30/climate-change-has-helped-bring-down-cultures/, Jan 30, 2012

[19] EPA, Climate impacts on agriculture and food supply, http://www.epa.gov/climatechange/impacts-adaptation/agriculture.html

[20] Strain, Daniel, *Warming dents corn and wheat yields*, Science News, vol. 179, no. 12, June 4, 2011, pg 15

[21] Telewa, Muliro, Kenyon drought means no school rations, BBC News Africa, http://www.bbc.co.uk/news/world-africa-12371130, Feb 8, 2011

[22] Gleick, Peter H, and Matthew Heberger, Devastating drought seems inevitable in American west, Sci. Am., http://www.scientificamerican.com/article.cfm?id=the-coming-mega-drought, Jan 5, 2012

[23] Than, Ker, Africans must adapt to drought in warming world: Report, Nat. Geo, http://news.nationalgeographic.com/news/2011/12/111216-africans-adapt-rainfall-shortages-warming-world/, Dec 16, 2011

[24] Parry, Wynne, Fierce heat waves and stronger storms coming, climate report warns, Live Sci., http://www.livescience.com/17111-extreme-weather-climate-change-report.html, Nov 18, 2011

[25] Rosen, Meghan, Tornado intensity climbing in the United States, Sci. News, Jan 25, 2014, pg 14

[26] Morello, Lauren, NOAA Chief: 2011 Weather was "Harbinger of Things to Come", Sci. Am., http://www.scientificamerican.com/article.cfm?id=noaa-chief-2011-was-harbi, Dec 8, 2011

[27] Live Science, Top 10 Surprising results of global warming,

http://www.livescience.com/11350-top-10-surprising-results-global-warming.html, Aug 16, 2011
[28] NASA, Climate change may bring big ecosystem changes, http://www.jpl.nasa.gov/news/news.cfm?release=2011-387&cid=release_2011-387&msource=11387&tr=y&auid=10010673, Dec 14, 2011

Chapter 2 –

[29] Castelvecchi, David, Degrees of Freedom, Fox Commentator Distorts Physics, http://blogs.scientificamerican.com/degrees-of-freedom/2011/08/11/fox-commentator-distorts-physics/, Aug 11, 2011
[30] Romm, Joe, Climate Progress, Accuweather's Joe Bastardi, denier of basic climate science, resigns after 32 years, http://thinkprogress.org/romm/2011/02/21/207573/accuweather-joe-bastardi-resigns/, Feb 21, 2011
[31] Cullen, Heidi, *The Weather of the Future, Heat Waves, Extreme Storms, and Other Scenes from a Climate-Changed Planet*, Harper, NY, pg 266
[32] *ibid.*, pg 30
[33] Somerville, Richard C. J. and Susan Jay Hassal, Communicating the science of climate change, *Physics Today, Vol 64*, No. 10, October 2011, pg 48
[34] True story
[35] Demelle, Brendan, Heartland Institute Exposed: Internal Documents Unmask Heart of Climate Denial Machine, Desmogblog.com, http://www.desmogblog.com/heartland-institute-exposed-internal-documents-unmask-heart-climate-denial-machine, Feb 14, 2012
[36] *ibid.* 22
[37] Nobel Prize-Winning Physicist Resigns Over Global Warming, http://www.foxnews.com/scitech/2011/09/14/nobel-prize-winning-physicist-resigns-from-top-physics-group-over-global/?test=latestnews
[38] Somerville, Richard C. J. and Susan Jay Hassal, Communicating the science of climate change, *Physics Today*,

Vol 64, No. 10, October 2011, pg 49

[39] Borenstein, Seth, 'Skeptic now agrees global warming is real', Associated Press, Oct 31, 2011, http://news.yahoo.com/skeptic-finds-now-agrees-global-warming-real-142616605.html

[40] NASA, Earth's energy budget remained out of balance despite unusually low solar activity, http://www.nasa.gov/topics/earth/features/energy-budget.html, Jan 30, 2012

[41] Stemweddel, Janet D, 'Evaluating scientific claims (or, do we have to take the scientist's word for it?)', http://blogs.scientificamerican.com/doing-good-science/2011/09/30/evaluating-scientific-claims-or-do-we-have-to-take-the-scientists-word-for-it/

[42] Pew Center on Global Climate Change, http://www.pewclimate.org/

[43] NASA Goddard Institute for Space Studies, http://www.giss.nasa.gov/

[44] NOAA Climate Services, http://www.climate.gov

[45] NASA Climate Change, http://climate.nasa.gov/

[46] Intergovernmental Panel on Climate Change, http://www.ipcc.ch/

[47] Galilei, Galileo, *Dialogues Concerning Two New Sciences*, Dover Publications Inc., New York, 1954

[48] Witze, Alexandra, 'Solar changes help create cold northern winters', Science News Web Edition, http://www.sciencenews.org/view/generic/id/335053/title/Solar_changes_help_create_cold_northern_winters, October 10, 2011

[49] Introduction to the scientific method, http://teacher.pas.rochester.edu/phy_labs/appendixe/appendixe.html

Chapter 3 – Is the planet warming?

[50] Ingham, Richard, *Is global warming causing harsher winters?*, Weather Underground, http://www.wunderground.com/news/global-warming-winters-20130328, March 28, 2013

[51] Powell, Devin, Less sea ice brings more snow, Science News, http://www.sciencenews.org/view/generic/id/338792/title/Less_sea_ice_brings_more_snow, Feb 27, 2012

[52] Morello, Loren, NOAA Chief: 2011 Weather Was "Harbinger of Things to Come", http://www.scientificamerican.com/article.cfm?id=noaa-chief-2011-was-harbi, December 8, 2011.

[53] Perry, Wynne, Fierce Heat Waves & Stronger Storms Coming, Climate Report Warns, http://www.livescience.com/17111-extreme-weather-climate-change-report.html, November 18, 2011

[54] Global temperature records in close agreement, http://climate.nasa.gov/news/index.cfm?FuseAction=ShowNews&NewsID=468

[55] *Ibid.*

[56] NASA Goddard Institute for Space Studies, GISS Surface Temperature Analysis, http://data.giss.nasa.gov/gistemp/graphs_v3/

[57] McKitrick, R. R., and P. J. Michaels, Quantifying the influence of anthrogenic surface processes and inhomogeneities on gridded global climate data, *J. Geo. Res, 112,* D24S09, doi:10.1029/2007JD008465, Dec 14, 2007

[58] Peterson, Thomas C., Kevin P. Gallo, Jay Lawrimore, Timothy W. Owen, Alex Huang, David A. McKittrick, Global rural temperature trends, Geophys. Res. Lett., vol. 26, no. 3, pp. 329-332, 1999, doi:10.1029/1998GL900322

[59] R. E. Benestad, R. E., Are temperature trends affected by economic activity? Comment on McKitrick & Michaels (2004), Clim. Res., Vol. 27: 171–173, Oct. 7, 2004

[60] Skeptical Science, Does *Urban Heat Island effect exaggerate global warming treands?*, http://www.skepticalscience.com/urban-heat-island-effect.htm

[61] JPL, New NASA web page sheds light on science of warming world, http://www.jpl.nasa.gov/news/news.php?release=2010-062

[62] Nongovernmental International Panel on Climate Change, *Acknowledging recent cooling,* http://www.nipccreport.org/articles/2010/jun/25jun2010a1.html

[63] Perlwitz, J., Hoerling, M., Eischeid, J., Xu, T. and Kumar, A. 2009. A strong bout of natural cooling in 2008. *Geophysical Research Letters* **36**: 10.1029/2009GL041188

[64] Cole, Steve and Leslie McCarthy, NASA Research Finds Last Decade was Warmest on Record, 2009 One of Warmest Years, NASA News Release 10-017, http://www.nasa.gov/home/hqnews/2010/jan/HQ_10-017_Warmest_temps.html

[65] Parry, Wynne, That's Hot! March Broke or Tied 7,755 U.S. Temp Records, Live Science, http://www.livescience.com/19448-march-temperature-record-high.html, April 3, 2012

[66] *ibid.* 57 Peterson

[67] NASA finds 2011 ninth warmest year on record, http://www.nasa.gov/home/hqnews/2012/jan/HQ_12-020_2011_Global_Temp.html, Jan 19, 2012

[68] Climate Change Reconsidered, Acknowledging Recent Natural Cooling, http://www.nipccreport.org/articles/2010/jun/25jun2010a1.html

[69] Why Satellite Temperature Data is more Accurate, http://theinconvenientskeptic.com/2010/10/why-satellite-temperature-data-is-more-accurate/

[70] Skeptical research effort confirms global warming, again, Morello, Lauren, and Climate Wire, Scientific American.com, October 21, 2011, http://www.scientificamerican.com/article.cfm?id=skeptical-research-effort-confirms-global-warming

[71] A new assessment of global warming, http://www.berkeleyearth.org/study.php

[72] NOAA, Arctic Theme Page, *Latest Surface Observations*, http://www.uni-koeln.de/math-nat-fak/geomet/meteo/winfos/synNNWWarctis.gif

[73] Remote Sensing Systems, Description of RSS/MSU and AMSU Data Products, http://www.ssmi.com/msu/msu_data_description.html

[74] Tropical lower tropospheric and surface anomalies from Jan 1958 to Dec 2010, http://www.metoffice.gov.uk/hadobs/hadat/images/update_imag

es/tropical_upper_air.png
[75] Climate Depot, http://www.climatedepot.com/
[76] Shine, Keith; M.S. Bourqui, P.M.D. Forster, S.H.E. Hare, U. Langematz, P. Braesicke, V. Grewe, M. Ponater, C. Schnadt, C.A. Smiths, J.D. Haighs, J. Austin, N. Butchart, D.T. Shindell, W.J. Randels, T. Nagashima, R.W. Portmann, S. Solomon, D.J. Seidel, J. Lanzante, S. Klein, V. Ramaswamy, and M.D. Schwarzkopf (2003). "A comparison of model-simulated trends in stratospheric temperatures". *Q. J. Royal Meteorol. Soc* **129** (590): 1565–1588. Bibcode 2003QJRMS.129.1565S. doi:10.1256/qj.02.186, http://pubs.giss.nasa.gov/docs/2003/2003_Shine_etal.pdf
[77] Satellite measurements of warming in the troposphere, http://www.skepticalscience.com/satellite-measurements-warming-troposphere.htm
[78] Heading towards the summer minimum ice extent, http://nsidc.org/icelights/2011/07/14/heading-towards-the-summer-minimum-ice-extent/
[79] Media Advisory: Arctic sea ice reaches lowest extent for 2011, http://nsidc.org/news/press/20110915_minimum.html
[80] National Snow and Ice Data Center, http://nsidc.org/cryosphere/sotc/sea_ice.html
[81] Solo Around the Americas, http://www.solotheamericas.org/
[82] PBS, Roald Amundsen, http://www.pbs.org/wgbh/amex/ice/peopleevents/pandeAMEX87.html
[83] NASA Goddard Space Flight Center, *ICESat and ICESat-2*, http://icesat.gsfc.nasa.gov/
[84] NASA Earth Science Project Office, *Operation Icebridge*, https://espo.nasa.gov/missions/oib/content/OIB_1
[85] ESA, CryoSat-2, http://www.esa.int/Our_Activities/Observing_the_Earth/The_Living_Planet_Programme/Earth_Explorers/CryoSat-2/ESA_s_ice_mission
[86] Gardner, Joan, Coincident Multiscale Estimates of Arctic Sea Ice Thickness, Eos, Transactions, vol. 93, no. 6, Feb 7, 2012, pg 57
[87] Weather Underground, *Permafrost in a warming world*,

http://www.wunderground.com/resources/climate/melting_perm afrost.asp

[88] ESA News, Signs of thawing permafrost seen from space, http://www.esa.int/esaCP/SEMHTAGY50H_index_0.html, Mar 27, 2012

[89] The University of Texas at Austin, *Coastal Antarctica permafrost melting faster than expected*, http://www.utexas.edu/news/2013/07/24/coastal-antarctic-permafrost-melting-faster-than-expected/

[90] The Engineering Toolbox – Water: Thermal Properties, http://www.engineeringtoolbox.com/water-thermal-properties-d_162.html

[91] Scientists excited by Arctic Ocean ridge finds, Mayall, Hillary, Nov 29, 2001, http://news.nationalgeographic.com/news/2001/11/1129_icebrea ker.html

[92] Mount St. Helens from the 1980 eruption to 2000, USGS Fact Sheet, http://pubs.usgs.gov/fs/2000/fs036-00/

[93] Joule to Megaton Converter Calculator, www.translatorscafe.com/cafe/units-converter/energy/calculator/

[94] NSIDC, *Patterns in Arctic weather and climate*, https://nsidc.org/cryosphere/arctic-meteorology/weather_climate_patterns.html#arctic_oscillation

[95] *ibid.*

[96] Iceberg Finder.comhttp://www.icebergfinder.com/iceberg-guide/iceberg-faq.aspx

[97] International Ice Patrol Homepage, http://www.navcen.uscg.gov/?pageName=IIPHome

[98] United Nations Climate Programme, Global Glacier Changes: Fact and Figures, pg 55, 2008 http://www.grid.unep.ch/glaciers/

[99] Ulmer, Simone, New calculations for the volume of Swiss glaciers, http://www.ethlife.ethz.ch/archive_articles/090622_gletschermas se_su/index_EN, June 22, 2009

[100] NASA, *Images of Change*, http://climate.nasa.gov/state_of_flux#Icemelt_Alaska8.jpg

[101] NASA Global Climate Change, http://climate.nasa.gov/

[102] NASA Mission takes Stock of Earth's Melting Land Ice,

http://www.jpl.nasa.gov/news/news.cfm?release=2012-036&cid=release_2012-036&msource=12036&tr=y&auid=10266783, Feb 8, 2012
[103] *ibid.*
[104] National Snow and Ice Data Center, State of the Cryosphere, http://nsidc.org/sotc/glacier_balance.html
[105] Levitus, S. J., I. Antonov, and T. P. Boyer, Warming of the world ocean, 1955 - 2003. Geophys. Res. Lett., 32, L02604, doi:10.1029/2004GL021592, 2005
[106] Hansen, J., L. Nazarenko, R. Ruedy, M. Sato, J. Willis, A. Del Genio, D. Koch, A. Lacis, K. Lo, S. Menon, T. Novakov, J. Perlwitz, G. Russell, G. A. Schmidt, and N. Tausnev, Earth's energy imbalance: Confirmation and implications. Science, 308, 1431-1435, 2005
[107] Lyman, J.M. and G. C. Johnson, Estimating Annual Global Upper Ocean Heat Content Anomalies Despite Irregular In Situ Ocean Sampling, Journal of Climate, 21, 5629-5641, doi:10.1175/2008JCLI2259.1, 2008
[108] Levitus, Syd, John Antonov, and Tim Boyer, Global ocean heat content 1955-2007 in light of recently revealed instrumentation problems, National Oceanographic Data Center – NOAA Ocean Climate Laboratory, CLIMAR III, May 2008, http://icoads.noaa.gov/climate3/c3oral-pdfs/S502-Levitus.pdf
[109] NOAA Ocean Climate Observation Program, Sea Surface Temperature, http://www.oco.noaa.gov/seaSurfaceTempProduct.html
[110] Lyman, J. M., J. K. Willis and G. C. Johnson, Recent cooling of the upper ocean, Geophysical Research Letters, 33, L18604, doi:10.1029/2006GL027033, 2006
[111] Willis, J. K., J. M. Lyman, G. C. Johnson, and J. Gilson, In situ data biases and recent ocean heat content variability. Journal of Atmospheric and Oceanic Technology, in press, doi:10.1175/2008JTECHO608.1, 2008
[112] NASA Climate Change – Key Indicators, http://climate.nasa.gov/keyIndicators/
[113] Perkins, Sid, Cooling climate 'consensus' of 1970s never was, Sci. News, vol 174, 9, Oct. 25, 2008, pg 5
[114] Struck, Doug, How the "Global Cooling" story came to be,

Sci. Am., http://www.scientificamerican.com/article/how-the-global-cooling-story-came-to-be/, Jan 10, 2014
[115] Record high temperatures far outpace record lows across U.S., http://www2.ucar.edu/news/1036/record-high-temperatures-far-outpace-record-lows-across-us
[116] Lindzen, Richard S., Deconstructing Global Warming, http://wattsupwiththat.files.wordpress.com/2009/10/cooler_heads_lindzen-talk-pdf.pdf, Oct 26, 2009.
[117] Romm, Joe, *Lindzen debunked again: New scientific study finds his paper downplaying causes of human-caused warming is "seriously in error"*, http://thinkprogress.org/romm/2010/01/11/205326/science-lindzen-debunked-again-positive-negative-feedbacks-clouds-tropics/, Jan 11, 2010
[118] Schmitt, Gavin and Michael Mann, *Lindzen in Newsweek*, RealClimate.org http://www.realclimate.org/index.php/archives/2007/04/lindzen-in-newsweek/, Apr 17, 2007
[119] *ibid.*
[120] True story. Lecture was at the U.S. Coast Guard Academy in the spring of 2007.
[121] Chapman, Phil, *Sorry to ruin the fun, but an Ice Age cometh*, The Australian, http://www.theaustralian.com.au/news/sorry-to-ruin-the-fun-but-an-ice-age-cometh/story-e6frg73o-1111116134873, Apr 23, 2008
[122] NASA, 2007 was tied as Earth's second warmest year, http://www.giss.nasa.gov/research/news/20080116/, Jan 16, 2008

Chapter 4 – Is the planet warming due to natural causes?

[123] NASA Mission News, Short-Term Ocean Cooling Suggests Global Warming 'Speed Bump', http://www.nasa.gov/vision/earth/lookingatearth/ocean-20060921.html, Nov 21, 2006
[124] The area of the world's oceans is 361 trillion (3.61×10^{14}) square meters. Of course, not all of the oceans are 2500 feet

271

deep. In fact, 2500 feet is pretty deep. So, I took half of that surface area to figure out the volume, or 1.8 x 10^14 square meters. 2500 feet equals 762 meters. Multiplying area by depth gives us a volume of about 1.37 x 10^17 cubic meters. It takes 4.184 x 10^6 joules to heat a cubic meter of water by one degree Celsius. But, it didn't heat that much, only .16 degrees Fahrenheit, which is equal to .089 degrees Celsius. So, multiplying the volume of water by the amount of heat needed to raise that water by one degree and then by the amount of degrees it was heated gives us a total amount of energy of about 5.1 x 10^22 joules of energy.

[125] Davies, J.H. and D.R. Davies, *Earth's surface heat flux*, Solid Earth, 1, 5–24, http://www.solid-earth.net/1/5/2010/se-1-5-2010.pdf, Jan 5, 2010

[126] Schwartzschild, Betram, *Neutrinos from Earth's interior measure the planet's radiogenic heating*, Physics Today, Vol 64, No. 9, pg 14, Sep 2011

[127] Roemmich, Dean, W. John Gould & John Gilson, 135 years of global ocean warming between the Challenger expedition and the Argo Programme, Nature Climate Change, doi:10.1038/nclimate1461, 01 April 2012

[128] Simkin T, Siebert L (2002-). Global Volcanism FAQs. Smithsonian Institution, Global Volcanism Program Digital Information Series, GVP-5 (http://www.volcano.si.edu/education/questions/).

[129] U.S. Energy Information Agency, International Energy Statistics, http://www.eia.gov/cfapps/ipdbproject/iedindex3.cfm?tid=44&pid=44&aid=1

[130] Science@NASA, Spotless Sun: Blankest Year of the Space Age, http://science.nasa.gov/headlines/y2008/30sep_blankyear.htm?list721778, September 30, 2008

[131] See, among others, Lockwood, M., R. Stamper and M. N. Wild, A doubling of the sun's coronal magnetic field during the last 100 years, Nature (doi:10.1038/20867), 399, 437-439, 1999; Solanki, S. K. and N.A. Krivova, Can solar variability explain global warming since 1970?, *J. Geophys. Res., 108*, 1200,

doi:10.1029/2002JA009753, 2003; Lockwood, Mike and Claus Fröhlich, Recent oppositely directed trends in solar climate forcings and the global mean surface air temperature, *Proc. R. Soc. A*, doi:10.1098/rspa.2007.1880, 2007; and Lockwood, M., and C. Fröhlich, Recent oppositely directed trends in solar climate forcings and the global mean surface air temperature. II. Different reconstructions of the total solar irradiance variation and dependence on response time scale, *Proc. R. Soc. A*, 464, 1367-1385, doi:10.1098/rspa.2007.0347, 2008

[132] Lockwood, M., and C. Fröhlich, Recent oppositely directed trends in solar climate forcings and the global mean surface air temperature. II. Different reconstructions of the total solar irradiance variation and dependence on response time scale, Proc. R. Soc. A 464, 1367-1385, doi:10.1098/rspa.2007.0347, 2008.

[133] Svensmark, Henrik, and Eigil Friis-Christensen, Variation of cosmic ray flux and global cloud coverage – a missing link in solar-climate relationship, *J. Atm. Sol-Terr Phys., Vol 59*, 11, July 1997, pg 1225-1232

[134] Lockwood, Mike and Claus Fröhlich, Recent oppositely directed trends in solar
climate forcings and the global mean surface air temperature, Proc. R. Soc. A
doi:10.1098/rspa.2007.1880, 2007.

[135] Brook, Ed, Paleoclimate: Windows on the greenhouse, Nature, Vol 453, 15 May 2008, p. 291-292.

[136] Usoskin, G, et al., Solar activity, cosmic rays, and Earth's temperature: A millennium-scale comparison, *JGR, 110*, A10102, doi:10.1029/2004JA010946, 2005

[137] Skeptical Science, Has the Sun been cooling or warming in recent decades?, http://www.skepticalscience.com/acrim-pmod-sun-getting-hotter.htm

[138] National Academies of Science, Climate Change at the National Academies, Figure 14: 800,000 years of temperature and carbon dioxide records, https://nas-sites.org/americasclimatechoices/more-resources-on-climate-change/climate-change-lines-of-evidence-booklet/evidence-impacts-and-choices-figure-gallery/figure-14

Chapter 5 – Is the planet warming due to manmade emissions?

[139] Marshall, Christa, and Climate Wire, Melting tundra releases carbon dioxide quickly, *Sci. Am.*,
http://www.scientificamerican.com/article/melting-tundra-releases-carbon-dioxide-quickly/, Feb 12, 2013
[140] GHG Online, Carbon Dioxide Sinks – Oceans,
http://www.ghgonline.org/co2sinkocean.htm
[141] NOAA, Earth System Research Laboratory, Trends in Atmospheric Carbon Dioxide,
http://www.esrl.noaa.gov/gmd/ccgg/trends/
[142] Gerlach, T.M., Present-day CO2 emissions from volcanoes: Transactions of the American Geophysical Union (EOS), v. 72, p. 249, and 254-255, 1991.
[143] *ibid.*
[144] *ibid.*
[145] Brantley, Susan L., and Kevin W. Koepenick, Measured carbon dioxide emissions from Oldoinyo Lengai and the skewed distribution of passive volcanic fluxes, Geology; October 1995; v. 23; no. 10; p. 933-936.
[146] Gerlach, T.M., Natural Sources of Greenhouse gases: CO2 emissions from volcanoes. Geothermal Resources Council Transactions, vol. 14, part I, pp. 639-641, 1990.
[147] Gerlach, T.M., Etna's Greenhouse Pump: Nature, 351, p. 352-353, 1991.
[148] Barker, John R., and Marc H. Ross, An introduction to global warming, *Am. J. Phys, 67(12),* December 1999.
[149] NOAA Paleoclimatology, The "Medieval Warm Period",
http://www.ncdc.noaa.gov/paleo/globalwarming/medieval.html
[150] Ljungqvist, Fredrik Charpentier; A new reconstruction of temperature variability in the extra-tropical northern hemisphere during the last two millennia, *Geogr. Ann.*, 92 A (3):339-351, 2010.

Chapter 5 – The Claims of the Skeptics

[151] NOAA, Paleo proxy data,
http://www.ncdc.noaa.gov/paleo/primer_proxy.html
[152] Encyclopedia Britannica, medieval warm period (MWP),
http://www.britannica.com/EBchecked/topic/175842/medieval-warm-period-MWP
[153] Rosenthal, Yair, et al., Pacific Ocean heat content during the past 10,000 years, *Sci, vol 342*, no. pp. 617-621
DOI: 10.1126/science.1240837, 1 Nov, 2013
[154] Schultz, Nora, Natural mechanism for medieval discovered, New Scientist, http://www.newscientist.com/article/dn16892-natural-mechanism-for-medieval-warming-discovered.html#.UxyM0s7NsdU, 02 April 2009
[155] National Snow and Ice Data Center, Average Monthly Arctic Sea Ice Extent September 1979-2013,
http://nsidc.org/arcticseaicenews/
[156] Jeffries, Martin O., James E. Overland and Donald K. Perovich, The arctic shifts to a new normal, Physics Today, Vol 66, No. 10, Oct 2013, pg 35
[157] Amos, Johnathan, Esa's Cryosat see Arctic sea ice volume bounce back, BBC News, Dec, 13, 2013,
http://www.bbc.co.uk/news/science-environment-25383373
[158] Fausto, Robert S., and William Colgan, Real time changes in Arctic ice presented in online portal, Eos, Vol. 94, No. 44, pg 397, 29 Oct 2013
[159] The Fram Museum, The Gjøa Expedition (1903 – 1906),
http://www.frammuseum.no/Polar-Expedition/The-Northwest-Passage-%281903-1906%29.aspx
[160] Historic Naval Ships Association, RCMPV St. Roche,
http://www.hnsa.org/ships/stroch.htm
[161] Solo the Americas, http://www.solotheamericas.org/
[162] KLRU, Nova, Arctic Passage,
http://www.pbs.org/wgbh/nova/arctic/
[163] The Week, Has global warming hit a plateau?,
http://news.yahoo.com/global-warming-hit-plateau-100000352.html, August 24, 2013.
[164] Global temperature records in close agreement,

http://climate.nasa.gov/news/index.cfm?FuseAction=ShowNews
&NewsID=

[165] *ibid.*

[166] Showstack, Randy, Climate report points to warming Earth, *Eos, Vol 94*, no. 34, pg 299, 20 Aug 2013

[167] Showstack, Randy, Decade of climate extremes, *Eos, Vol 94*, No. 29, pg 255, 16 July 2013

[168] *ibid.* 159 The Week

[169] Balcerak, Ernie, and Colin Schultz, Past decade saw unprecedented warming in the deep ocean, Eos, Vol 94, No. 32, pg 288, 6 Aug 2013

[170] Skeptical Science, What were climate scientists predicting in the 1970s?, http://www.skepticalscience.com/ice-age-predictions-in-1970s.htm

[171] Perkins, Sid, Cooling climate 'consensus' of 1970s never was, https://www.sciencenews.org/article/cooling-climate-%E2%80%98consensus%E2%80%99-1970s-never-was, Sci. News, Oct 25, 2008

[172] Struck, Doug, How the "Global Cooling" story came to be, *Sci. Am.*, http://www.scientificamerican.com/article/how-the-global-cooling-story-came-to-be/, Jan 10, 2014

[173] Painting, Rob, Climate models show remarkable agreement with recent surface warming, SkepticalScience.com, http://www.skepticalscience.com/Climate-Models-Show-Remarkable-Agreement-with-Recent-Surface-Warming.html, March 28, 2014

[174] Rose, David, *Global warming stopped 16 years ago, reveals Met Office report quietly released ... and here is the chart to prove it*, Mail Online, http://www.dailymail.co.uk/sciencetech/article-2217286/Global-warming-stopped-16-years-ago-reveals-Met-Office-report-quietly-released--chart-prove-it.html, 16 October 2012

[175] Milne, Richard, Climate skepticism or denial? The battle to inform public opionion, The University of Edinburgh, http://blogs.sps.ed.ac.uk/global-environment-society-academy/2013/07/26/skeptics/, July 26, 2013

[176] McKitrick, Ross, R., and Patrick J. Michaels, *Quantifying the influence of anthropogenic surface processes and*

inhomogeneities on gridded global climate data,
DOI: 10.1029/2007JD008465, *J. Geophys. Res., Vol. 112*, Issue
D24, 27 December 2007

[177] *ibid.* 156

[178] Vincent, Warwick F, Mickael Lemay and Michel Allard;
Adapting to permafrost change: A science framework, Eos, Vol.
94, No. 42, pg 373, 15 Oct 2013

[179] Cook, John, Does Urban Heat Island effect exaggerate global
warming trends?, http://www.skepticalscience.com/urban-heat-
island-effect-intermediate.htm

[180] Campo et al., Independent confirmation of global land
warming without the use of station temperatures,
doi: 10.1002/grl.50425, *GRL, Vol 40*, Issue 12, pg 3170, 28 June
2013

[181] Schult, Colin, Independent observations corroborate surface
air temperature record, doi: 10.1002/2013EO350012, *Eos, Vol
94*, Issue 35, page 312, 27 August 2013

[182] Mann, Michael, The Hockey Stick and the ClimateWars.
Dispatches from the Front Lines, Columbia University Press,
2012, pg 68

[183] Skeptical Science, Lindzen's London Illusions,
http://www.skepticalscience.com/lindzen-london-illusions.html,
7 March 2012

[184] This lecture was attended by the author at the U.S. Coast
Guard Academy in the spring of 2008.

[185] Roy Spencer (scientist),
http://en.wikipedia.org/wiki/Roy_Spencer_%28scientist%29

[186] *ibid.* 182, pg 24

[187] *ibid.* 182, pg 182-183

[188] Climate misinformer: Roy Spencer,
http://www.skepticalscience.com/skeptic_Roy_Spencer.htm

[189] Examining Dr. John Christy's global warming skepticism,
http://www.skepticalscience.com/examining-christys-
skepticism.html

[190] Desmogblog.com, http://www.desmogblog.com/anthony-
watts

[191] *ibid.* 182, pg 72

[192] Fox News, Scientists call AP report on global warming

'hysteria', http://www.foxnews.com/story/2008/12/16/scientists-call-ap-report-on-global-warming-hysteria/, Dec 16, 2008
[193] Associated Press, Obama left with little time to curb global warming, http://www.oregonlive.com/environment/index.ssf/2008/12/obama_left_with_little_time_to.html, Dec 15, 2008
[194] NOAA, http://www.ncdc.noaa.gov/img/climate/globalwarming/ar4-fig-ts-18.gif
[195] Increased hurricane activity linked to sea surface warming, http://www.sciencedaily.com/releases/2008/01/080130130647.htm, Sci. News, Dec 31, 2008
[196] Mann, Michael E., *The Hockey Stick and the Climate Wars: Dispatches from the Front Lines*, Columbia University Press, NY, 2012, pg 207
[197] *ibid.*, pg 208
[198] *ibid.*, pg 67
[199] Negin, Elliot, Unreliable sources 5: How the news media helped the Kochs and ExxonMobil spread climate disinformation, http://www.huffingtonpost.com/elliott-negin/unreliable-sources-5-how_b_3316714.html, May 22, 2013
[200] Los Angeles Times, Climate denial in the classroom, http://articles.latimes.com/2012/feb/20/opinion/la-ed-climate-20120220, Feb 20, 2012
[201] Revkin, Andrew, More on Peter Gleick and the Heartland files, Dot Earth, http://dotearth.blogs.nytimes.com/2012/02/22/more-on-peter-gleick-and-the-heartland-files/?_r=0, Feb 22, 2012

Chapter 7 – What Can Be Done?

[202] Carrington, Damian, *More than 1000 new coal plants planned worldwide, figures show*, http://www.theguardian.com/environment/2012/nov/20/coal-plants-world-resources-institute, The Guardian, 19 Nov 2012
[203] IPCC, Climate Change 2001: Mitigation, B. Metz et al., Eds., http://www.grida.no/publications/other/ipcc_tar/?src=/climate/ip

cc_tar/wg3/index.htm

[204] Hoffert, M. I., et al., Advanced technology path to global climate stability: Energy for a greenhouse planet, *Science* 298, 981 (2002)

[205] List of countries by carbon dioxide emission per capita, http://en.wikipedia.org/wiki/List_of_countries_by_carbon_dioxide_emissions_per_capita

[206] Pacala, S. and R. Socolow, Stabilization wedges: Solving the climate problem for the next 50 years with current technologies, *Science, vol. 305*, pg 968, 13 Aug 2004

[207] http://www.whitehouse.gov/the-press-office/2012/08/28/obama-administration-finalizes-historic-545-mpg-fuel-efficiency-standard

[208] http://www.bloomberg.com/news/2011-11-16/doubling-fuel-economy-may-cost-2-000-a-car-157-billion-u-s-draft-says.html

[209] U.S. EPA, Draft Inventory of U.S. Greenhouse Gas Emissions and Sinks, 1990 – 2012, Executive Summary, http://www.epa.gov/climatechange/ghgemissions/usinventoryreport.html, Feb 2014

[210] This meeting occurred between the author and General Motors engineers and occurred at the U.S. Naval Academy in the spring of 2011.

[211] Environmental Defense Fund, *How cap and trade works*, http://www.edf.org/climate/how-cap-and-trade-works

[212] Broder, John M., *'Cap and trade' loses its standing as energy policy of choice*, http://www.nytimes.com/2010/03/26/science/earth/26climate.html?_r=0, the NY Times, Mar 25, 2010

[213] *ibid.*

[214] Ecomii, Cap and trade, http://www.ecomii.com/ecopedia/cap-and-trade

[215] Kolbert, Elizabeth, *Paying for it*, http://www.newyorker.com/talk/comment/2012/12/10/121210taco_talk_kolbert, The New Yorker, Dec 10, 2012

[216] ExxonMobil, ExxonMobil's views and principles on policies to manage long-term risk from climate change, http://corporate.exxonmobil.com/en/current-issues/climate-policy/climate-policy-principles/overview?parentId=dfdb24e1-

090d-4900-9e84-1b931453fc66

[217] Harvey, Fiona, Climate change is already damaging global economy, report finds, http://www.theguardian.com/environment/2012/sep/26/climate-change-damaging-global-economy, The Guardian, Sep 25, 2012

www.ingramcontent.com/pod-product-compliance
Lightning Source LLC
Chambersburg PA
CBHW051210170526
45166CB00005B/1832

* 9 7 8 1 5 0 0 2 1 0 0 5 2 *